THE FINAL TIDE

THE FINAL TIDE

Norma Cole

Jesse Stuart Foundation
ASHLAND, KENTUCKY
1999

The Final Tide

Copyright © 1999 by Norma Cole

All rights reserved. No part of this book may be reproduced or utilized in any form or by any means, electronic or mechanical, including photocopying, recording, or by an information storage or retrieval system, without permission in writing from the Publisher.

Library of Congress Cataloging-in-Publication Data

Cole, Norma.
    The final tide / by Norma Cole.
       p.   cm.
    ISBN 0-945084-71-4
    Summary : When the Tennessee Valley Authority builds a dam at Wolf Creek to bring electricity to Tollers Ridge, Kentucky, everyone in fourteen-year-old Geneva's family prepares to move to higher ground except for Granny who refuses to leave her home.
    [1. Tennessee Valley Authority--Fiction. 2. Moving, Household--Fiction. 3. Grandmothers--Fiction. 4. Kentucky--Fiction.]
    I. Title.
PZ7.C673526Fi
[Fic]--dc21
                                                                       99-12402
                                                                           CIP

Published By:
The Jesse Stuart Foundation
P.O. Box 391 Ashland, KY 41114
(606) 329-5232 or 5233

TO MY MOTHER, EDITH

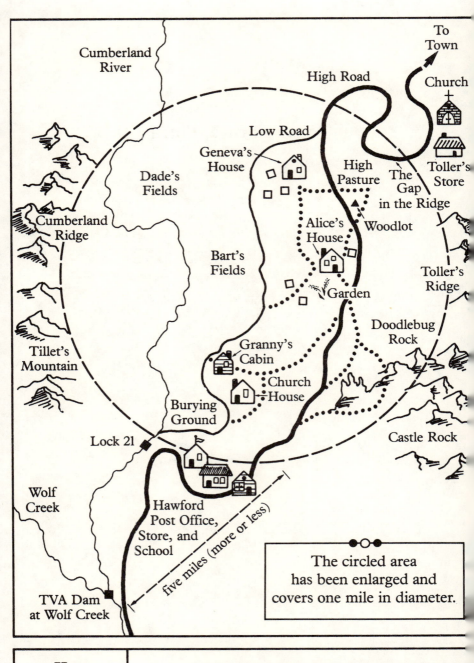

1

FOR THE THIRD DAY, RAIN RATTLED against the tin roof. Inside, the house was shadowy and chilled, though both stoves carried a good fire. Geneva's mother trimmed the wicks and filled the lamps and lanterns with coal oil. Geneva worked with her mother, cleaning the thin glass globes with a soft cloth. It was their regular Saturday-afternoon chore.

Geneva heard her daddy on the porch before her mother noticed. When he opened the back door, she saw her mother's back stiffen. I reckon Mamma's worried about the tide, Geneva thought. Not this tide, which would go down in a few days, but the final tide, the one that would take everything of their life and leave nothing familiar to begin with again.

"I need the gal," Dade said to his wife. He put the lantern on the table and took a filled one.

Geneva looked to her mother. Mattie gave a slight nod.

Dade smiled at Geneva. "Tide's rising. You take Suky and lead the cows up to high pasture. Hurry along before dark. Can you hold old Bossy on another rope? You know how skitterish she is."

He turned to Mattie. "Milking's done, ma'am. On the table fer yer separating. I done it early on account of the tide. Now, Geneva, mind where you put yer feet. Them snakes'll be leaving the pasture fields an' woodlots, too. That's good, wear my old jacket an' yer boots. Get something on yer head. Rain's cold."

Geneva scrambled to follow his directions. She needed few; she had been through enough tides to know what to do. Yet this was the first time her daddy had asked her to take care of the cows.

Dade lit his lantern. Mattie worked on the one he had put on the table. They spoke quietly until they saw Geneva. Dade shook his head. "You ain't growed into my old jumper yet, child. Turn up them sleeves. Yer hands'll need to be free with that old Bossy to lead."

"Chickens above the tide?" Mattie asked.

Dade nodded. "Better see to them turkeys, though. Turkeys seem to get dumber every year. I'm working on the brood sows and the sheeps to do next, though I think their pasture'll be above the water. Can't tell how much more rain we'll get."

Geneva preceded Dade through the kitchen, out the door, and across the porch. Once the door had shut, he said, "Mark the day well, young'n, this'll likely be the last tide before the water gets to us. Next tide'll not go down in a day er two."

Geneva imagined how, before bed, she would write the date in her notebook: March 6, 1948. A Saturday.

By this third day of rain the Cumberland River had swollen and overflowed its banks and water lay across their rich bottom lands. When the water passed and the spring sun had warmed and dried the fields, her daddy and Uncle Bart would plow and plant and tend them for the last time.

She'd heard the dam at Wolf Creek was almost finished. When the dam was done, the river would no longer wind its way between Toller's Ridge to the east and the mountains to the west. Its water would be stopped to make a mighty lake. The next tide would cover her world. She and Mamma and Daddy would move to Toller's Ridge.

Geneva and her father followed the well-trod path down the hill from the house to the barns and shed below. She looked back at the familiar shape of her house printed against the lowering sky and the darker hills beyond. Wispy smoke clung to the two chimneys. A soft lamp-glow outlined the kitchen window. Mamma would be starting their supper soon, though they would not eat until all the animals were safe.

"Think the tide'll get the house?" She was not really worried. She could remember plenty of tides, and her home had always been above the floodwaters.

"The house'll be fine through this one, but the next one'll get it all." Geneva knew no words to comfort her saddened father.

She looked beyond the house and around the curve of the hill. She hoped to see Alice and her cows moving along their path toward the high pasture. If Alice was out, the best-friend cousins would meet and walk home together. Alice's mother had kept Alice so busy Geneva hadn't seen her all day. A day without Alice is a day without sunshine, Geneva thought. She chuckled. It truly had been a day without sunshine.

Alice was nowhere to be seen. Geneva sighed. She would have to struggle alone up the hill and down again. "Geneva!" Dade's voice was sharp. "Wake up, child. I'm thinking yer mamma may be right. Maybe it is time to make a lady out of you. You sure ain't much of a hired farmhand lately."

The girl chose to think her daddy was teasing. "I'd rather be a no-good hired farmhand than a lady any day of the year, Daddy. So please don't fire me yet."

"Then let's see if you can get that Suky in one hand and manage the old Bossy with the other an' get the rest of them to tag along. I reckon I can't fire you till they's up to the high ground anyway."

She led the gentle Suky and the obstinate old Bossy through the mud of the cow yard and along the path to the high pasture. Once started, the rest of the cows followed quietly. At the high pasture, Uncle Bart's cattle lowed and called with the pleasure of the meeting. "Drat," Geneva muttered. "I wish I was meeting Alice too.

I know! she thought. I'll run down there before I go home. If I hurry, Mamma'll not notice I went.

She took the path around the woodlot, unmindful

of the dusky dark. A misty fog, replacing the driven rain, swirled upward. Good, she thought, mist's rising. Rain's nearly over. I'll tell Mamma that, an' how dark it was, when she wonders what took me so long.

Geneva knew she should go directly home, but for a few minutes visiting with Alice, the younger cousins, and her cheerful Aunt Ina she would risk her mother's anger.

She met Kendall, Alice's next younger brother, outside their barn. "You come to play marbles?" he sang out. "Reckon we could get us up a game this fine spring evening?"

Geneva knew this wasn't entirely teasing. She had won his favorite marbles the last game they'd played before winter, and he rankled about it all the time. "You get the lantern, boy, I got yer marbles," she retorted.

Not for the first time Geneva wished Kendall were her brother rather than her cousin. She was sure her life would be easier if there were brothers to share the attention paid to her at home.

Oh, there had been a brother once, though she did not really remember. Neither Mattie nor Dade ever spoke of that baby; all Geneva knew was that he was buried up in a little patch garden her mamma tended and made pretty. Geneva seldom followed the path to this high ground, though her mother found her way up there often.

Kendall broke the silence. "Race you to the house." Geneva let him win, though she knew her longer legs could have outrun his.

"Child, what you doing here?" Aunt Ina asked.

"Come to see if yer all right. Where's Alice?" Geneva shook the water off the brim of her large straw hat.

"She's rocking the baby while I get supper on. Will you eat with us, child?"

"No, ma'am. Mamma's getting ours on the table, I reckon."

Alice's singing voice came from the bedroom just beyond the kitchen table. "I'm in her-r-re. Shhhhh-shhhh-shhhha."

Geneva, aware of her muddy boots, stopped in the doorway. She could hardly see Alice in the rocking chair, swaying back and forth, but she could hear the movements of the fussy baby. "I ain't seen you all day. You all right?"

"Yes, I am fine," Alice said in a tune, "but you better not disturb this fussy thing 'cause I've been rocking her forever an' ever."

"What's wrong with Sissy?"

Alice sang her reply. "Teeth, teeth, teeth, the baby's getting teeth. The sister rocks the baby, but the baby does not close her eyes. Teeth, teeth, teeth."

Geneva laughed at the song. "I'll see you tomorrow. Reckon we'll have church to home. Otherwise we'll have to go to the church house in a boat."

Geneva turned back to the warm, lamp-lit kitchen. "Anyone been down to see about Granny Haw today?" she asked.

"I didn't go," Ina said, looking to the circle of children. "Did yer daddy go? Did any you young'ns?" A chorus of "no" met her questions.

"Then I'll quick run down there." She opened the back door. "If my daddy comes looking fer me, tell him I'll be right back."

She skittered along the muddy path in the near dark. "I'm glad I know my way," she said, feeling foolish for being out without a lantern and for talking to herself, a habit about which she was often teased.

Despite the familiarity of the path, she found herself sitting in the mud from a misplaced step, and she was well along the path to Hawford Store and Post Office before she realized she'd missed the turn to Granny's in the dusky wetness.

Finally the church house loomed in the gray light and the expanse of water below brought her the full extent of the tide. Lordy, she thought, I better get Granny to go home with me. Much more rain an' her cabin'll be an island. Yet the girl was sure Granny's house, built high on a rock foundation, was not in danger.

"Gran-n-y. Granny!" Geneva shouted; she wanted to alert the old woman that someone was outside. The front door opened and her grandmother paused in a frame of light from her lamp.

"That you, child?"

"It's me, Granny. I come to see if yer all right. I come to take you home with me."

"Law, child, I ain't going home with you. I got to

stay here with my treasures. I got to keep the tide out a my cellar. I got no time to be nowheres but here. Anyway, if the tide was to take me it'd be my Lord's will, an' I could just lay me down by my man, there, an' be with him in our heavenly home."

"Granny!" Geneva scolded, not sure what the old woman seemed to be saying. "We don't want the tide to get you. Come on, now, come home with me."

"No, I won't. You just get on out a here. You ain't even got you a lantern, child. Your mamma know yer off down here without even a light?"

"No, actually, not really. It's okay, though. I wanted to be sure you were safe."

"Child, I'm an old woman an' I can take care of myself. Here, come in here an' get you a lantern and get you on home. I'm a lot safer than you are being out in this wet dark. No telling what's out there to get you."

Geneva followed the old woman into the kitchen, where Granny quickly took a lantern from the shelf. "I got to watch my treasures." She struck a match and in its glow gave Geneva a searching look. "You know about my treasures, don't you, child?"

Geneva didn't, but she nodded her head. "Sure, Granny. I understand."

"Then you know I can't go home with you. Can't let no water into my cellar or get my 'taters an' such. But still, my treasures are sealed in blue jars. Reckon they'd be safe. No. I can't go off an' leave them to the tide. Worst come to worst, child, an' I'll take us into the loft, there. If the tide gets that deep you an' Bart's an' all'd be gone, too.

"I'll stay to home. You run now, child. Yer mamma an' yer daddy'll be fussing about you."

Geneva stepped off the porch into total darkness. Guilt overwhelmed her; she had been gone from the house far too long. There'd be trouble waiting for her at home. A gust of wind swung through the trees, dripping on Geneva, who shook with a sudden chill. From the porch Granny called: "Wind's up, child. Storms does that now an' then. Blowing the clouds off, I reckon. Hurry, now."

Granny closed the door. Grateful for the lantern, Geneva plodded up the rise she had, minutes before, slid down. The next gust of wind brought a driving rain and sent her large straw hat careening into the dark. "Damn." She caught the forbidden word: "Drat. Now I'll get wet inside as well as out." Rain sluiced behind the too-big jumper collar and down her neck. She pulled the collar under her chin and, with her free hand, wound her long braids around the outside.

She made her way up the high path, choosing her steps, for it was littered with downed tree branches, slippery rocks, and puddles of standing water. Rain, powerful in a new burst of fury, soaked her already-damp pant legs. A cold drizzle slithered into her boots. "There's not an inch a dryness on me," she grumbled. "Won't I ever get home?"

Alice's house looked cozy in the dark and wet, the

windows lamp-lit and beckoning. She longed to stop. There would be no scolding from her aunt or uncle about her late and wet escapade to Granny's. They would all welcome her, helping her off with the wet clothes, bringing dry ones and hot coffee. But a different fate awaited her at her own house and Geneva hurried to meet it.

Far ahead, she saw a faint lantern glow bobbing along the path to the high pasture. "Gen-e-va! Geneva!" Her daddy's voice wavered above the noise of the wind and the rain.

"Here I am, Daddy." She stumbled in her hurry, wanting to assure him she was safe and to feel his presence.

"What you coming from that way for?" he yelled.

"I went down to see if Granny was all right," she called into the space separating them. "Nobody'd been to see to her today."

Dade's light and voice circled the house and closed the space between them. "Why'd you go off without telling your mamma? She's plumb wore out with fretting about what's happened to you up on the mountain. Child! When will you ever learn to . . . here, give me that lantern. Get on home." With a burst of speed Geneva scooted across the porch to meet her mamma in the open back door.

Mattie's face first showed relief, next concern, then anger. "Geneva Augusta Haw! Where have you been?"

Geneva unwrapped her braids and hung her head.

Guilt overwhelmed her, but a wave of defiance quickly pushed it aside. She had done nothing wrong. Well, she *should* have told her mamma she was going to see about Granny, and she'd actually gone first to see Alice. She'd done that because she'd wanted to; concern for her grandmother came next.

Dade followed Geneva through the still-open back door. "Gal, get out a them wet clothes," he ordered. Seeing a way out of her predicament, Geneva responded with a loud, fake sneeze.

Mattie took up the sneeze. "Yer wet through an' through. Child, you ain't had but one cold this winter. Now yer asking fer it. Get that jumper off. Where's yer hat? There ain't a dry place on you."

Geneva shivered, weariness and cold replacing her guilt and defiance. Mattie brought an old quilt. With a pile of wet outer clothing at her feet, Geneva pulled it around her shoulders and dropped the rest of her clothes behind the wrap. "Get over to that fire," Mattie directed. "Yer daddy's waiting on his supper an' I'm fooling with a thoughtless child. Where'd you go beside that high pasture?"

"I went to see if Granny was safe." She used a small voice.

"An' I suppose that included a stop at Ina's? Sometimes I think we ought to pack you up an' send you down there to live." Geneva had never been able to tell her mother that she liked watching and listening to the loud and boisterous cousins but preferred her own quiet home. Mattie brought her a cup of coffee well laced with cream and sugar. "Get dry clothes on now, child. We'll eat supper and then we'll see what's to be done about you."

Geneva escaped up the steps to her room. Her mamma had not yet forgiven her, and later there'd be a punishment to face.

Supper, as usual, was mostly quiet. "Mist's rising from the mountains." Geneva ventured a break in the chill.

"Rain's almost over," Dade answered. Mattie said nothing. Geneva brought them each a cup of coffee as was their habit and began to clear the table.

"Do the dishes, then go to bed," Mattie directed.

Was this to be her punishment? Sent to bed? If so, Geneva could only be glad, for weariness filled every bit of her body. She would miss the Saturday night radio shows, but if she could stay awake, she'd likely hear most of them from her room.

Dade moved from the table to his easy chair. "Let's see if the battery'll pull in the radio through the rain." He connected a car battery to the radio with the wires he'd rigged up. Silence followed the click of the on/off button. "Gotta warm her up," he muttered, and in response the radio sent a wave of static into the quiet kitchen. He moved the dial. Static and screeches and snatches of voices repeated until a bit of music caught his attention. "Almost time fer 'Renfro Valley Barn Dance.' You gonna send the child to bed?" he said to Mattie.

"She's going to bed. No radio show fer her. She might get a whipping."

Geneva responded with a series of sneezes. The only thing that could change her mother's mind was the threat of sickness. Mattie rose to the bait, went to the bedroom, and brought a soft cloth to warm by the stove. Geneva immediately knew what that meant and was sorry for the fake sneezes.

Mattie dug into the onion bin for already-sprouting onions. She began chopping them furiously. She dropped them into a skillet of hot grease and added a generous helping of dry mustard. "I'll put you to bed with this poultice on yer chest," she said, "and we'll get this thing before it gets started. You know how you are with cold in yer chest. We got enough on our minds without fussing with no sick child."

"I don't need no mustard plaster, Mamma. I'll be all right. I'll just go to bed now. Rest'll do me more good than anything."

"You get into them nightclothes an' get on down here. We'll do this thing now an' we'll do it right."

An hour later Geneva, sweating and reeking of mustard and onion, lay in her bed. "Renfro Valley Barn Dance" music floated around her ears and steam from the hot mustard and onions kept tears in her eyes. I would rather have had the whipping, she thought. I'm in this onion mess all because I'd thought to fool my mamma. Serves me right, she concluded.

2

THE RAIN SLACKENED THROUGH THE night. By dawn, halfhearted clouds let themselves be driven off by a mild spring sun. Geneva wondered, as she stretched and yawned, if this would be an at-home church day.

Whichever it was, the sounds from the rest of the house let her know it was time to rise. She slipped into old overalls, outdoor shoes, and a sweater in time to hear Mattie's voice at the foot of the stairs.

"Geneva, yer daddy's already gone to the barn."

This was the second year Geneva had taken her mother's place at the milking. Mattie's hands had gotten crippled with rheumatism and she found it too painful to use them the long minutes it took to wring milk from the cows.

Geneva clattered down the steps.

"Child! Remember you are a lady, despite them overalls."

"Are the cows down?" Geneva could hardly believe that they were.

"No. Yer daddy's gone to hitch up the wagon.

You'll do yer milking in the high pasture."

The girl wolfed down a chunk of leftover corn bread and made her way out the back door. Leaving the privy, she met Dade bringing mule and wagon into the yard. "Morning, young'n," Dade shouted. "It's a fine one. Rain's over. Let's go."

Geneva was surprised at his cheer; he had become so silent and glum of late.

She sprang to the wagon seat. "We going to take Uncle Bart an' the boys?"

"If them lazybones er ready. I'll stop on the high road an' you can run in to see."

Uncle Bart and the younger boys, Kendall and James, were gathering pails and stools when Geneva crossed the yard. "We're on the high road. Alice coming too?"

Bart shook his head. "She's helping Ina with them babies. The little one's cranky with the teething, I reckon. Yer daddy's on time, I see." Bart lifted his milk can into the wagon. It clanged against theirs. "Morning," Bart said to Dade. His brother gave a slight nod.

Kendall and James climbed into the back with Geneva. Uncle Bart pulled himself up and onto the seat by his brother. Geneva listened to the space between her daddy and her uncle, the brothers Haw. It would tell her how the day would go.

"Would you look at that pretty water," Bart exclaimed.

"Next water'll not be so pretty," Dade grumbled.

"No, not to us it won't." Bart's voice was cheerful. "But it sure will be pretty water to go through the dam an' make electricity fer folks all over this part of Kentucky. Why, them fellers up there in Washington, D.C. knew we needed to get into the twentieth century 'stead a living old-timey forever."

"Yer speechifying," Dade said. "Far as yer concerned, it don't matter we're losing everything."

"Dade, we got no right to hang on to some old farm when our land'll give people 'roundabout a chance to live better."

Geneva saw her daddy's back grow rigid. "Yer own brother died in that there Normandy invasion not five years ago. Why'd you think he died and got buried there in France?"

"It was World War Two," Bart explained. "He was serving his country same as I was there in Detroit working in that defense factory. I was building them B-Fifty-two bombers and he died from a German bullet."

Dade bristled. "Just like you to have a smart answer. I'll tell you why he died an' I'll tell you why you worked in that city. He died so I could live free. Well, living free means living here without no gov-ment telling me I gotta move an' change. Now, yer a different case. You worked in that city to get yer head full of big ideas."

"Now *yer* speechifying," Bart interrupted.

Geneva listened to the men. They had had this argument many times, but she still couldn't rightly decide which one she agreed with.

She'd tried to imagine the dam at Wolf Creek as it had been described to her many times. "They say them gov-ment men filled a place between two mountains full of concrete an' dirt. Is that true?" she'd asked her daddy.

"That's what I hear tell. I hear they got big machines set in that concrete. Turbines, they call them. An' when that squeezed-up water pours through them turbines, electric comes pouring out. I can't hardly believe it neither," Dade had answered.

Geneva knew about electric, though they had none on their own farm nor did anybody in the valley. Up in town there was electric, street lights, even, though it seemed a waste to light up streets when everybody was to home. She'd heard people made a stove hot with electric and cooked that way. It would be nice if she and Mamma didn't have to make a fire to cook, especially in the summer.

Geneva remembered a moving picture she'd once seen up town. Someone had said that a camera turned around and around and a light went through to send the picture on the wall where all the people were looking. She knew electric could make things cold, too, like ice cream and such, but Geneva seldom got to go to town, so she'd never been in a person's home to see for herself.

People up town had telephones, too. Did those need electric? Daddy's radio ran on batteries, so maybe telephones did, too.

Geneva looked across the shiny, sparkling water of the tide to the mountains beyond. She tried to think how the space between any two of them could be filled up with concrete and dirt, but there was no way she could imagine the dam at Wolf Creek.

Nor could she imagine the lake that would be covering their farm with six hundred feet of water. Why, the lake would reach almost to the tops of the mountains!

Dade's voice brought her back to the jogging ride to the pasture. "Don't matter to you that Haws have lived here since Daniel Boone showed them the way. You don't care yer people always made their living farming an' taking care of their selves an' their land."

"Sure I care, Dade. But there's more to do than live down here. You work this land without a tractor. You can get you a tractor now, what with the money the gov-ment settled on yer land."

"Hmmmmph. Tractor don't make up fer the lives a all them Haws that lived before us. Anyway, them mules got to earn their keep."

"No, but them old Haws worked hard to get something better an' now you can get something better without working so hard. You've known about this dam since 1935. Why, it's been fifteen years, nearly, they told us the Tennessee Valley Authority was gonna build them dams all along the river here an' in Tennessee. People been talking forever about them electric-making dams. Don't know why yer so against it now the time is here."

"I guess I never believed them." Dade's voice was soft. Bart was silent.

"You just want to quit farming an' yer too scared to tell me," Dade said.

"You got to admit, Dade, them kids'll be better off in town. They'll get them a good education an' live easier than we done. You best think to send Geneva to us an' let her go to high school, too. That young'n might could go to college."

"She don't need to get above her raising."

"I sure want to go to school with Alice an' the boys . . ." Geneva let her voice trail off, wishing with all her power her daddy would finally agree.

Dade turned and frowned. "Geneva, hush. It's been settled." She dropped her eyes. If Uncle Bart couldn't make him see reason, she'd never get to go.

Her thoughts drifted to the school she'd never get to go to. She'd heard the girls up there wore twin sweater sets that matched their skirts, and those saddle shoes. She's seen them in the catalog, brown and white and so pretty, but her mamma said white shoes in the winter was silly.

Alice told her kids at school had dances *together*—boys and girls both! Even if she could go to high school she'd never be allowed to go to a dance with boys—not that she wanted to, really, but it would be a sight to see. Alice told her kids had records and record players that played music. She remembered an old victrola with a horn sticking out the top. It belonged to her Great Aunt Dorcas. The music was so funny and scratchy nobody could ever have danced to it.

Alice said kids in town didn't start school till nearly September instead of July like they did. And they didn't get out of school till May! Why, her country school closed down at Christmas. Did those town kids walk to school in the snow and rain and ice all winter?

Course, they didn't have spring farm work to do, so they might as well be in school, she decided. Well, I got spring farm work to do. Reckon that's one reason Daddy'll not let me go.

Alice said when she got to town she was going to get her hair cut and get it curled up with a permanent. Geneva sighed. Lucky Alice.

Kendall pulled at her arm. "Hey, Geneva, wake up. What you thinking about, anyway?"

"I was thinking about that dam at Wolf Creek. What you suppose that thing looks like, anyway?"

"I saw a picture of it in a magazine I was looking at up town. You recollect that day we went to Lock Twenty-one with Daddy?"

Geneva nodded. Lock Twenty-one was downriver from the Hawford Store and Post Office, down where the river had cut a deep gorge. Uncle Bart was lock keeper, and he'd shown the children how he was to open three iron gates in the concrete dam and let the water out of the river. Or he closed the gates, when it was dry, to keep water in.

"I think the dam looks like Lock Twenty-one,"

Kendall said. "Only really big. Big enough to fill up the space between any two mountains you look at right now."

"I reckon yer right."

"We ought to get yer daddy an' mine to take us down there," James said, "an' see them men building that thing. My daddy'd do it, would yers?"

"I don't think so." Geneva sighed. "Seems like my daddy thinks it all ain't gonna happen if he pretends the dam ain't down there." He's just plain stubborn about cooperating with them gov-ment men, she thought.

Geneva knew well where her daddy got the stubborns. He was almost as stubborn as Granny. At least he was fixing to go. Geneva thought about her Granny, who refused outright; "My man is buried right there an' I ain't going an' you ain't taking his grave away from here, neither. Them gov-ment men ain't gonna chase off an old woman. I'll shoot 'em if they tries." All of them—Dade, Mattie and Bart—knew Granny meant to do just what she said.

Mattie and Ina often talked about their mother-in-law in sad, regretful words. "Oh, she was a wonder in her day," Ina would say: "Why, when Bart took me home to meet her, she was as bright an' cheerful a woman as you ever met! An' work! My, how that woman could work!"

"An' willful," Mattie would add. "We knows where them Haw men get their stubborns, don't we?"

"Seems like after Granddaddy Haw passed on an'

that boy a hers didn't come back from the war and that spell took her, seems like all's left of that woman is the stubborns."

"Well," Mattie would say, "I can't do nothing with her nor can you. Reckon it'll be Dade an' Bart who'll have to solve her problem."

"She could come to town with us," Ina would agree, "an' I know she could go up on the Ridge with you. But I reckon it'll kill her if we move her."

"An' I reckon it'll kill her if we leave her," Mattie said. Their talk moved on but the worry stayed with Geneva, for it was her granny she admired most.

At the high pasture Kendall staked his first cow next to Geneva's. If her cousin Alice was her best friend, Kendall was what she imagined as the perfect brother. Only a little more than a year younger than Alice, Geneva heard plenty of complaints about him. But, for a boy, Geneva considered him the best.

"You coming down this afternoon?" Kendall kept his voice low. "You got my taw an' I want it back."

Geneva grinned. Kendall was almost the champion marble player of Hawford school, but she had his best marble, his pretty taw. She'd had it since their last game, way back in November. If Mamma knew she was planning to still play marbles, Mamma would have a fit. At the very least, Mattie would take away the bags of well-earned marbles, and Aunt Ina would be told about Alice playing, too.

Not that it wasn't all right for girls to play marbles now and then when they were young, but Geneva and

Alice were fourteen, long past the age when squatting in the dust and competing with boys was acceptable.

"Think we'll go to church?" Geneva answered. "If we do we'll have to go by boat."

"I heard Ma tell Pa we could have church down home an' yer Ma was to see to dinner. I guess the women already decided."

"Then they'll sit around the house an' talk." Geneva grinned with satisfaction. "An us kids'll go behind yer house and I'll get yer shooter as well as yer favorite taw."

"No siree-bob," Kendall protested. "I'll get yer shooter an' my taw back, an' yer taw, too."

"I aim to beat all the boys in yer grade once school takes up again." Geneva knew how to tease Kendall. "You an' that smarty Selby Lynch, an' Bosford. They won't be expecting any girl to do it, but I will."

Kendall shrugged. He knows I will, too, Geneva decided.

Kendall carried his pail of milk to the wagon, poured it into the milk can, and brought another cow to the place where he and Geneva milked. James, who had favored a place near his father and his uncle, came by with his first full pail. "You gonna give Kendall a chance to get his taw back? That's his bestus marble, you know. You had it all winter an' it's a-burning him."

Geneva and the boys grinned. It was settled. Their Sunday afternoon of play would begin with marbles, the first game of the spring.

3

"WE'LL HAVE CHURCH DOWN AT INA'S." Mattie served their breakfast. "Just 'cause there's a tide don't mean we can't do right by the Lord. Then they'll come here for dinner. So, Geneva, there's lots to be done between now an' dinner time. Don't you plan on getting yer nose into a book er getting away from me."

"Yes, ma'am."

"You think what hymns you want to play this morning an' take time fer a little practice. I reckon my fingers er getting so bad you'll take over the church playing when the Brother comes back first Sunday."

Geneva was not surprised at this announcement, she had been expecting it. Mattie had always played the reed organ at church and Dade led the singing, but the rheumatism in Mattie's hands had made music playing more and more difficult. Geneva suspected part of the idea was to give her more grown-up responsibilities, to make her a lady.

Mattie had given Geneva all the organ lessons she'd ever had, and Geneva knew she played well.

Their little church house down by Granny Haw's shared a preacher with three other churches. That meant the Brother was with them every fourth Sunday.

Whether the Brother was with them or not, there would be Sunday school, singing and praying. At the church house, if one of the men felt moved to speak, he would do so.

Church at home would be little different. They would have Sunday school, women and girls in one group, men and boys in the other. But in the privacy of their home, Mattie or Ina might speak her feelings.

The girls, followed by Alice's sister Betty Lou, escaped to Geneva's room as soon as dinner and the dishes were finished. "Go away, Betty Lou," Alice scolded. "Go play with Tad."

"Aw, come on, Alice. Betty Lou's okay." Geneva said.

"I know she's okay," Alice complained, "just a pest. I take care of her all the time, I can at least have Sunday to myself."

"I won't bother you none," Betty Lou said, "an' if you don't let me come, I'll tell Mamma an Aunt Mattie what yer fixing to do."

"That's blackmail." Geneva laughed, but she knew Betty Lou had won her point.

"Put yer overalls in a poke-bag. How about a book? A big book that looks squarish. Here, open the poke. Hide yer marbles behind that," Alice directed. "Put on a blouse. How about an old skirt? An' a sweater. You can tell yer mamma yer chilly. You can put yer overalls on over home."

"Alice! Yer rushing me," Geneva complained, but she hurried to do as Alice had said. She distributed her marbles to several pockets of the overalls and folded them into the poke. She picked up a hated crochet hook. "I know, I'll let some of this stupid thread stick out of the sack. I won't lie, but Mamma'll think we're going to crochet."

"I'll ask if I can borrow her new pattern book," Alice added. Geneva envied, sort of, Alice's mastery of the dratted hook. It pleased the older women that Alice made even stitches and pretty patterns. "Let's go. You better come, Betty Lou, an' don't you be letting nothing out."

The three girls clattered down the steps. "Walk, girls, walk," Mattie chided. "Remember you are fourteen and must be more ladylike, an example for Betty Lou."

"Yes, ma'am," Alice murmured.

"We're going to Alice's," Geneva called, rushing through the kitchen where the women, Tad—Alice's four-year-old brother—and the baby were visiting. Uncle Bart, Dade, and James were on the front porch, relishing the spring sun.

Alice paused. "Can I borrow yer new pattern book, Aunt Mattie?"

Geneva chuckled. That Alice, she's a real faker.

"Why, of course, dear. You send it on home with

Geneva at supper time. I'm grateful my rheumatism ain't keeping me from my crochet work. I'm making a ruffle out a that book an' will want it back this evening."

Geneva, listening outside, smiled at how easily Alice had fooled Mattie.

"Thank you, Auntie." Alice, Betty Lou, and Geneva started down the path.

"What'er you gonna do?" James yelled from the porch. "Wait fer me."

"Don't get into any meanness," Bart called after them. His laughter followed them into the quiet Sunday air.

Below the barn the spring sun and the drainage of the river were already seeping the tide off the pastures and fields. The children hardly noticed the water: It had arrived with such fury and fright, but its leaving was quiet.

Kendall waited on the front porch. He had his last bag of marbles clutched in one hand and his favorite lucky shooter in the other. He rubbed the shooter against a rabbit's foot. Geneva, with a look at the boy's face, knew he was not sure his good-luck magic would help. "I swept off the shooting ground," he said. "It's pretty damp. Ought to make fer good shooting."

The children ran to the play area. With no windows on that side of the house and with the smokehouse at the other end, it seemed far away from prying eyes. The third side of the space went steeply up the hill to the high road and the fourth opened into the family garden where the soil was already turned and ready for the first planting of early peas and onions.

Geneva drew a circle in the middle of the smooth earth. Kendall drew a larger circle around the inner one. "Come on, Kendall, that's too close," Geneva protested.

"No it ain't. James wants to play. When we knock him out, we can move it out."

Geneva lay down three empty stockings and one filled with marbles. Stockings made good marble bags, she thought, an' I'll give everybody but Kendall his marbles back.

"First shot to James," Kendall said. The children nodded. James knelt to take his shot, but he missed by a wide margin. Soon he was out of the game. Kendall put Alice out next.

"You'll give me my marbles back, won't you, Kendall?" Alice's voice was full of sisterly sweetness.

"Naw."

"You will too or I'll tell Mamma."

Kendall grinned. "No you won't. You ain't supposed to be playing no more. Yer a girl, an' too big."

"I'll tell Daddy, then. He don't care if I play marbles. Mamma don't either. It's just Aunt Mattie who tells her I shouldn't."

Geneva sighed. Mattie had gotten these notions

about the time of her fourteenth birthday and it was spoiling all the fun. Oh, it was kind of nice to be considered grown-up when they had company or went somewhere. She was allowed to stay with the women and help with the baby children. Otherwise, it was a pain in the neck to be constantly reminded what girls did or did not do.

Kendall patted his almost-marble-empty pockets. "I ain't gonna give nothing back until I get mine. When I win you might-maybe get yers."

"Come on, Kendall. Play or quit," Geneva said.

He crouched to study his shot. Geneva bent over him and let her long braids switch gently around his ears.

"Quit it. Yer hair's tickling me," he complained.

"That's my secret weapon." But she flipped the long braids over her shoulder.

"Secret weapon nothing," he muttered. "Cheating I calls it."

He picked up a marble and held it toward the light. "It's nearly a tigereye," he said. "I'd sure like it back."

"You can win it fair an' square if yer good enough," Geneva teased.

Kendall gave her a sideways look. "How about a game of all-or-nothing?"

"Why not?" Geneva dropped all the favored marbles into the circle.

"You better draw lots for the first shot," Alice said.

She broke a twig into two uneven parts. She held one in each hand behind her back and shuffled them between her hands.

"You sure you don't know which is which?" Kendall accused. "It'd be just like you to give the short one to her."

"I mixed them up," Alice said. "You draw first if yer thinking I cheat."

The children paused as Kendall studied Alice's arms. Into the silence charged Mattie and Ina. "What's going on here?" Mattie's voice thundered.

Geneva jumped, Betty Lou screamed, Alice dropped her arms. Kendall dropped his hand to the circle and swept up his favorite taw.

"You put that back there," Geneva said. "That's mine an' you know it."

"How long has this been going on?" Mattie demanded.

The children fell into a line facing aunt and mother. Mattie's face was a study of anger. Ina's held a sparkle of humor. Geneva knew who was in trouble now, and it wasn't any of her cousins.

"Geneva, have you been playing marbles?"

"Yes, ma'am."

"I thought we had decided you were not to play anymore. We agreed to that after yer birthday."

"Yes, ma'am."

"Kendall, has she taken yer marbles?"

"Yes, ma'am, Aunt Mattie."

"All of them?"

"Nearly all, ma'am."

Mattie picked up three marble-filled stockings and handed them to the boy.

"Hey, one of them's mine," James protested.

Though one of the stockings held Alice's marbles, she said nothing.

"Which one?" Mattie asked James. He silently accepted the stocking Kendall handed over.

"Kendall, you may keep the rest of them. Geneva'll never need them no more." Mattie turned to the girl. "Young lady, you march on home an' go to yer room. I'll take care of you later." Mattie turned and preceded Geneva out of the yard. Geneva walked backward. She pointed from Kendall to Alice, and the boy handed his sister the marbles she'd lost. He mouthed a silent ha-ha-ha.

Geneva stuck out her tongue. Behind her mother she took the long ends of her braids and crossed them under her nose as though they were a long, droopy mustache. Betty Lou and James giggled.

"Geneva!" Mattie commanded.

The last thing Geneva saw was Kendall scoop up her shooter where it lay on the knuckle line and kiss the back of his hand.

I'll see about that, Geneva thought; I'll get back my marbles an' my shooter an' yer marbles too, Mr. Smarty-pants. You know the rules same as me.

4

MATTIE STOOD IN GENEVA'S DOORWAY. "You will spend this evening in yer room, an' you will think about the decision we made an' what you done today. You will read these passages in the Bible an' be able to tell them to me in the morning. They're about a child obeying his parents."

"Yes, ma'am."

"Yer too big fer a whipping, but if this happens again, a whipping you'll get."

"Yes, ma'am."

"An you'll not go to Alice's fer playing fer one whole week. In fact, next Sunday we'll decide if you'll go next Sunday."

This was the cruelest blow of all. She would be stuck with only Mattie and Dade for company. But Geneva knew to answer:

"Yes, ma'am."

She also knew she would not stop playing marbles. If Mamma wanted to test her share of Haw stubborns, Geneva was prepared to show them. Maybe

Mamma wouldn't know Geneva was playing marbles, but she *would* play.

Mattie turned. She walked, very ladylike, down the steps.

Geneva stood up. She closed the door with a good hard slam but caught it before it crashed. She flopped on the bed. She threw her arms across her chest and frowned. There were no tears. This is not worth a tear, Geneva thought, though she felt a sting behind her eyes.

Actually, she wasn't all that angry or surprised. In fact, she had expected the whipping. Maybe there were some advantages to growing up.

Geneva opened the door. "How about the milking this evening?" she called to the empty stairs.

"I'll do it," Mattie answered.

"But what about yer hands an' the rheumatism?"

"Fine time fer you to be thinking about my hands an' my pain."

Guilt overwhelmed Geneva. She knew of no other girl who would give her mamma such pain.

Geneva sat at her table. She pulled the Bible to her. She found the passages Mamma had showed her and read them diligently. They concerned boys, mostly; at least they talked about "he." Well, she wasn't any boy. It didn't apply to her. Without the slightest thought to the meaning of the words, she committed them to memory. Remembering things was easy for her, and after a few minutes, she found the book she had been reading, opened it up, and began to loosen the braids of her hair. Except for being hungry, she rather enjoyed her quiet evening.

"Geneva, milking time!" Mattie's voice awakened her from a pleasant dream, a dream about getting her hair cut and fixed like the girls in the catalog. Geneva touched her head. The long stuff was still attached.

After breakfast, Geneva carried buckets of water into the kitchen where Mattie had the copper boiler on the stove and a good fire going. "While I get this water hot," Mattie directed, "you run down to Granny's an' get her dirty clothes. Take that pillow slip, the old one, an' bring 'em back in that. She won't have much, but if she ain't stripped her bed yet, you help her with it. You can stay and talk to her fer a minute. She'll be wanting to tell about the tide, I reckon, an' maybe there's something we need to do fer her."

Geneva gathered up the pillow slip. "Mamma, can I take her a piece of the cake you made yesterday?"

"That'll be fine. She likes her sweets. Tell her yer daddy'll be fetching her up here tomorrow so's we can get her bathed an' her hair washed. She won't let me help her down there an' she ain't got water handy in the pump like us."

"Yes, ma'am."

"Don't you be stopping at Alice's hear? Yer being punished an' I'm trusting you to do like I say."

"Yes, ma'am."

"I'd send you on the low road but fer the tide in the lower pasture, so you better go the high one. Mind what I said about Alice."

Geneva escaped.

Alice and Betty Lou were already at the clothesline. Alice hung the wet clothing as Betty Lou handed her clothespins. Uncle Bart had brought home a real gasoline-fired washing machine when he came back from Detroit and his war work, so washing was fast at their house.

Geneva paused on the high road above the house. "Howdy. I ain't supposed to talk to you."

"I reckoned that'd be the way it'd be," Alice answered. "Well, I ain't supposed to talk to you neither. Did you get a whipping?"

"No. Only I can't play with you fer the whole of the week an' maybe not next Sunday either."

"That's what my mamma said, too. So how come yer talking to me now?"

"I'm going to tell on you," Betty Lou sang out.

"You better not, er I'll get you," Alice said.

"Aw, can't you take a joke? Geneva knew I was joking, didn't you, Gen?"

Geneva nodded. "Let's leave notes in the secret place."

"Okay. I'll leave one this evening an' you can answer in the morning. I can get it easy."

"Let's plan on three a day, or maybe five."

"Like we done last time we was punished," Alice agreed.

"You better send me one, too," Betty Lou cut in. "I get to have notes, too."

"Well, you gotta answer if you do an' you ain't very good at writing," Alice scolded.

"Oh. I can make pictures."

"I'd like yer pictures," Geneva said, "an' I'll answer you back with pictures."

"Alice? Betty Lou?" Aunt Ina rounded the corner of the house. "Here, girls, these are ready." She put down a another basket of wet clothes.

"Bye," Geneva said. "Morning, Auntie. Nice morning, ain't it? Tide's going down. I'm on my way to Granny Haw's. Can I tell her yer news?"

"Tell her Sissy finally got that tooth an' she ain't so fussy. Now, Geneva, I know you ain't to be talking to Alice, so you go on now, 'fore you get in bigger trouble."

"Yes, ma'am."

From the far end of Ina's garden, Geneva could see Granny's cabin. A wisp of smoke from her morning breakfast fire drifted into the early morning sunshine. Next door, the church house and cemetery lay peaceful and calm above the retreating water. Geneva flung her arms over her head, let out a wild war whoop and scissor-legged down the incline, plummeting ever faster until she coasted into Granny's yard breathless and filled with laughter.

Granny Haw emerged from the front door. Geneva looked with wonder at the nearly toothless old woman, tiny to the point of elfin, worn slippers peeking below a long aproned skirt, dust cap covering her long hair, and eyes as bright and as black as coal.

"Come on in, gal. Had yer breakfast yet?"

"Could always use another biscuit an' some molasses." Geneva knew this would please the old woman.

"Come in. Come in, don't be standing there. I got two–three left. Might could give you a wee drap a coffee."

"I'd like that." Geneva followed her through the main room of the small cabin into the lean-to kitchen.

Granny set out the plate of almost-warm biscuits, the jar of molasses, and a cup of coffee. Geneva reached into the crock where Granny kept the silverware and brought out a knife. Granny perched on a chair. Her birdlike eyes darted from Geneva's hand spreading molasses to the mouth waiting for the first bite of feather-light biscuits.

"I was sitting here thinking about my treasures," Granny began. Oh, oh, Geneva thought, sometimes she talks as loony as a tune an' this must be one of those times.

"You knows I got me treasures in heaven an' treasures here on this earth an' treasures in blue jars. Mind you, now, gal. There's treasures in blue jars."

Geneva wondered if she should try to humor the old woman, contradict her, or listen carefully. Perhaps she could find out what Granny meant. Granny often talked about treasures in blue jars, but only to Geneva or the other children. Neither Mattie nor Dade had ever heard such talk.

Granny seemed to forget her thought. She continued. "Wasn't that a fine tide? Oh, it do me good to see them waters rising. Some folks fear the tides, but not yer Granny Haw. Yer Granddaddy Haw, he respected them tides. Mind me, now, when I die I'm to lie beside him in the place in the churchyard. I'll go soon, gal, soon. He allas said the tide brought new dirt, new treasures fer the earth. Some a them treasures er in blue jars, gal. Remember that."

What is she trying to tell? Geneva wondered. "I come to get yer wash, an' Daddy'll be over to get you tomorrow fer a bath an' a hair washing."

Granny pulled off her old-timey dust cap. Long, still-black-and-white hair tumbled out and settled around her shoulders. "Why, Granny, you ain't braided yer hair yet this morning." Geneva was surprised. In the daytime, Granny's hair was always up.

"I'm tired, tired, tired of this stuff. I'm gonna cut it off."

"Oh, Granny, would you really do that? You ain't been to town for ever, but would you get it cut off if you could? Have you ever had a haircut? I haven't!"

"Tired, tired, tired. Don't need this hair. It ain't a treasure. Gonna cut it off."

"I've a mind to do it, too," Geneva agreed, thinking to humor the rambling talk.

"You do it, child. You cut that burden off. Be free of it, be free. Get you an education. Make you a teacher. Get it off."

Fat chance, Geneva thought. Daddy won't even let me go to high school.

"Come on, Granny. Mamma wants me to help you change yer bed so she can do yer sheets. You ain't done that yet, have you?"

"Yer mamma's right good to me, gal. She's been a real care, fer a daughter-in-law. She's a might grim, that one is, an' that other one's flighty. Why, hang on to her and you'll end up no telling where."

Granny's reference to Ina brought her message to mind. "Aunt Ina says to tell you Sissy finally got her new tooth an' ain't so fussy now. She thanks you fer the herbs that raised it out."

Granny laughed. "That makes Sissy with more teeth than me."

Geneva breathed a sigh. Granny seemed to be back in this world. "Wouldn't it be fine if you could raise you some teeth with them herbs?"

"I done tried, girl. I tried me every yarb in the woods for teething babies' an I didn't get nary one."

In the big room, Geneva began pulling quilts off Granny's bed. Granny fluttered her hands among the bedding in a large chest. In a moment Geneva was down to Granny's shuck mattress. "Granny, wouldn't you like Daddy to bring you a real mattress? Seems like yer old-time shucks would be uncomfortable an' yer old bones deserves a nice, new store-bought mattress."

"Don't need none. Them shucks is plenty good enough, though I could use me a few more in it now I've squashed them all winter. Don't bother, though, I ain't sleeping there long. I'll soon be in my heavenly home, having my e-ternal rest with yer granddaddy."

"Aw, Granny, yer strong an' healthy. You ain't going to yer heavenly home fer years an' years."

"Child, I ain't asking you fer lies. I'm talking serious here."

Geneva heard a note of pleading in Granny's voice. She knew how she hated it when her mamma or daddy wouldn't talk about real things. "I'm sorry. I won't tease you, Granny. What do you want to talk about?"

Granny shook a clean, white sheet over the bed. Geneva caught the other side and together they stretched it out across the lumpy shuck mattress.

"I'm talking about dying, child. That was my last tide. Afore the next one, I'll be gone."

"But, Granny, the next tide is likely the for-ever tide. They got the dam at Wolf Creek all done. Come the next tide, they'll be catching the water in a lake near as big as the ocean."

"I ain't gonna be here fer that tide." Granny flung out her feather-filled comforter. "An' that's what I want you to understand. You gotta be the one to make yer daddy understand. An' that Bart. Yer daddy can't see beyond the end of his nose an' that Bart can't even see his nose end. Between them, they don't know nothing but farming an' men's business."

Geneva chuckled in agreement with Granny's view of her daddy and her uncle. "You don't have to die to stay out of that tide. You can come live up on Toller's Ridge with us. I'd like that."

"No, I ain't leaving here. An' neither is yer granddaddy. Nobody is gonna move his grave, hear? Them other graves can go wherever the gov-ment takes 'em, but my man stays here. Him an' me got us a promise an' I aim to keep my side of it."

Granny flung a quilt on top of her sheets. Geneva had never seen the old woman move so fast. "We ain't moving. I'm counting on you to make them leave me alone."

"But Granny, I'm just a child. Grown-ups don't listen to me." Geneva caught the edge of the next quilt.

"They'll listen to you when you tell them about the treasures in blue jars."

Oh, oh, here we go again. "Is that what you want me to tell them, Granny, that you won't leave because you have treasures in blue jars? Are you sure that's what you want me to say?" They'll think I'm as daft as you if I do, Geneva thought.

Granny laughed. "No, child. No. Them treasures

is fer you young'ns. Now, you gotta think a something to make them listen to reason. I gotta depend on you." Granny flung out another quilt. "Neither of them'll listen to me. Why, it'd be just like yer dad to come over here with that mule an that wagon an a big rope an—an—kidnap me!"

The picture of her father and Granny caught in the act of kidnapping was too much for Geneva. She burst into laughter. In a moment Granny joined in.

Granny tossed another quilt out of the chest. With little thought, Geneva opened it and flung it out. Granny smoothed it down on her side of the bed. "Seriously, Granny, I don't know what I can do about getting them to listen."

"Just hold them off. Make them give me time to . . . just don't let them—put them off. You tell 'em *later. Not yet. Maybe next week.*'

"You can do that. You can make them leave me alone till I die. I promise you, I won't be long. Not near long. My treasures er waiting in my heavenly home."

Geneva sighed. "I'll do what I can, Granny. But I can't promise."

"Oh yes you can promise. You ain't leaving here till you make a promise an' a solemn vow."

Geneva wondered if Granny were teasing, but the look in the black eyes told her Granny was perfectly serious.

Granny flung another quilt. Geneva caught it and

smoothed it down. "You must promise." Granny rounded the end of the bed and clutched Geneva's arm. Her small, sharp fingers pressed into Geneva's skin.

"Yer really serious about this, ain't you, Granny?"

"Promise." Granny's fingers bit her flesh. Geneva couldn't believe the old woman's strength.

"I'll . . . I'll try."

"Promise."

There seemed to be no other way. "I promise." I promise to do the best I can, Geneva thought, but I likely won't succeed.

Granny let go of Geneva's arm. She turned, bent over the chest, and threw another quilt on the bed.

Geneva absently brushed out the wrinkles, then realized what they had been doing. "Granny! How many quilts are you going to use? You must have fifteen on here."

Granny peered into the chest. One last quilt lay folded in the bottom. The shuck bed was piled high with quilts of every pattern, combination of color, and age.

Granny giggled. "Oh, child. I'm such a ninny. I done put nearly every quilt I own on my bed. What'll they do with an old silly-head like me?"

Geneva picked up the top quilt to fold. "Never mind, child. I got the whole day to right this mess. While I got them out, I'll make some decisions about them."

"I'd best go home, then. You know how Mamma is about washing the white clothes first. She'll be waiting on yers."

Geneva picked up the pillow slip with pretended heaviness and made Granny laugh. Granny followed Geneva across the front porch. "Oh, you be a funny one, ain't you? You put me in mind of me when I was yer age an' I was already married. Yer but a slip of a girl. You get that education, now, an' make you a teacher. Don't be slipping off with no boy when you ain't but a child."

Beyond the fence Geneva turned to wave. How small she is, Geneva thought, how small and daft and stubborn.

"Remember what I said," Granny called. "Remember yer promise. Remember yer treasures in blue jars. Remember there are treasures on earth an' treasures to store in heaven, but don't wait too long fer the treasures in yer heavenly home. Have yer treasures here, too."

Geneva waved again. She was surprised at the strength of Granny's voice, clearly audible in the soft, early-spring air.

"An' make you a teacher. I'm a-going one a these days. Just like that tide out there in the field today, I'm a-going."

5

THE WEEK'S FUN WAS PASSING NOTES TO Alice and exchanging pictures with Betty Lou.

Geneva's note of Tuesday posed a riddle to Alice:

> *Wait until you see Granny. She came for her bath yesterday. She took ACTION. Why can't we? Go to see her.*

Alice answered:

> *Come on. Tell. I can't go to Granny's right now.*

Geneva:

> *No.*

Alice:

> *I'll get even with you.*

Geneva:

> *There is a clue in the picture I made for Betty Lou.*

Alice:

> *It looks like a picture of a place where they do hair.*

45

Geneva:
You are a good detective.

By Thursday morning Geneva decided Alice was having her revenge.

She couldn't tell if Alice had found out about Granny's hair or not.

The moment Granny came for her bath on Tuesday she pulled off the dust cap she usually wore and her cut-off hair stood out in every direction. Mattie gasped. After a moment, Geneva laughed. "So, Granny. You did it."

Granny giggled. "I ain't never been so free. You do yers, gal. You too, Mattie. It's old-timey to have that old long stuff hanging around. I wish I'd done it years ago. Don't I look a sight?"

Mattie could say nothing. While Granny bathed, Mattie turned to Geneva. "You seem to know something about this." It was a statement more than a question, but Geneva felt the need to explain.

"She's talked about it fer a long time. She's getting so daft an' says so many things, I can't tell what she means an' what she don't."

"I see. I suppose that's why you didn't tell me or yer daddy."

Geneva couldn't understand her feeling of guilt. Did her mother think she could have stopped Granny had she known? It wasn't exactly any of her mother's business, Geneva concluded and remembered

Granny's description of her mother as "rather grim." Geneva decided she certainly was.

"Is there anything else she's been saying that you've kept from yer daddy er me?"

"I don't know. She talks so silly so much of the time, I can't tell. She talks about treasures in heaven an' treasures here on earth an' treasures in blue jars, but that sounds completely daft to me, don't it to you?"

"Yes. Anything else?"

"Well, she don't want to go to Toller's Ridge with us, or to town with Uncle Bart. She wants to be buried near Granddaddy."

"We know that."

"She wants me to see to it that neither Daddy nor Uncle Bart takes her away by force, but I don't know how I can do anything about their decision."

"No, I don't think you can." Mattie was thoughtful, Geneva decided, not angry. "Well, time will tell."

By Thursday evening Alice had, apparently, found out about Granny's hair and it was her turn to lure Geneva into a speculation:

Ask your daddy about the big news for Sunday.

Geneva wondered how she could ask questions of Dade without pointing to the fact that she was getting information from Alice. She talked to him at milking time. "Daddy, is the tide down enough to have church Sunday?"

"Reckon it is. Folks er coming down from Toller's Ridge to meet with us an' see if we want to join up with them. 'Course, Bart an' Ina er going to town, but I reckon us an' most of the neighbors going to the Ridge'll join."

"Folks from the Ridge, huh. How many?"

"Don't rightly know, but we'll eat on the ground after service an' get to know everybody better. Course, we know them people, but we ain't never considered if we want to church together an' they'll want to know more about us down here."

Geneva's note of Friday read:

> *Work on your mamma. See if we can be together when we eat on the ground after church.*

Alice answered:

> *Mamma don't care. She says Mattie makes these decisions.*

As Geneva and her folks arrived at the church house, Mattie asked Geneva to fetch her granny. "See if you can't get some kind of a hat er a bonnet on her head. She'll do it fer you."

"Then can I find Alice?"

"Don't reckon I can keep you from it 'less I tie you to my belt. Yer too big fer that an' there's strangers here today."

Geneva sprinted toward Granny's. "Walk!" Mattie commanded. Geneva adopted a ladylike gait. "That's more like it," she figured her mamma was saying, and,

"Whatever is to become of that child?"

Granny was still in the big room. "Why, Granny, you look right nice this morning." Geneva viewed the old woman with interest. Granny had on her usual long skirt and a bright new apron. "Are you going to wear that to church today?" Geneva indicated the apron.

Granny glanced down. "Why, ain't I a silly? I meant to roll that up into my poke. We're eating on the ground after church an' I'll want to help."

"Why don't you wear that pretty new spring hat we got you out of the catalog?" Geneva asked.

"Don't want to. Want folks to see my new fine haircut. Maybe some a them other women'd take a hint."

"But, Granny, have you looked at the hat with yer new hair?" Geneva held out the black straw hat with its puff of fake flowers and bit of veil tied around the back.

Granny set it on her head. She looked this way and that in her faded mirror. "What you think, child?"

"I like it."

"Well, let's go. This is a morning to enjoy. A treasure on earth."

After the service the women spread big blankets under the trees. Mounds of food were piled on a cloth-covered wagon bed and clean buckets of lemonade floated the last of the winter's ice. It had been decided that each Ridge family would sit with a river-bottom family. They had known each other all of their lives, but with the three or four miles between river bottom and ridge, such an occasion was rare.

Once they had eaten, the children raced around and in and out of the groups. Alice and Geneva sat with the girls, all of whom had been reminded that it was Sunday and that they were nearly grown-up.

Geneva looked at the girls who sat by her, the ones she had known all her life from school and home and the strangers from Toller's Ridge. They were all all right, she supposed, but there was not one she really wanted for a friend like Alice. Yet she had to make a friend, for Alice would be moving to town and a different world. "Oh, Alice," Geneva moaned, "I think it's mean of yer daddy to take y'all off to town."

"I feel bad too, Gen, but I'm excited. He says I'll get to work in the store when Mamma don't need me. Don't you worry, though. You'll still be my best friend and my cousin. Nothing can change that! Why, you'll get to come to stay with me, an' I'll get to come to you."

Geneva sighed.

She looked at the younger girls who ran and played with the boys. Where was the cutoff? When was a girl considered too old to play? The boys never seemed to be chided about their games. There were Kendall and James in the thick of mumblety-peg, and Kendall was only a year younger than Alice. Geneva looked for the boys her age or older. They were in the field having a game of town ball. Surely they were "playing."

She saw Kendall and James and several boys head for the back of the church house. A marble game would soon be under way. Oh, how she longed to play. But she couldn't, not under her mother's watchful eye, and besides, Kendall had her blue taw and her other shooter and all the rest of her marbles.

In a moment she saw her mother pull out the new crochet book, the hook, and the thread, ever present in her poke. "Come on, Alice," she whispered. Geneva nodded toward the women, where Mattie had become the center of attention. Alice quickly followed Geneva out of sight and range of their mothers.

"What's going on?" Alice asked.

"The boys have gone behind the church house. Bet they're playing marbles back there on the sand."

"Are you fixing to play?"

"Well, no—I guess. I'd like to, but I reckon I better not chance it."

"Good. I sure know I won't."

The girls strolled around among the trees and the groups of visitors. By this time, all the men were sitting on the church steps. They talked of crops and the news of the dam at Wolf Creek and speculated as to how long it would take for the last tide to cover the river bottom. After Geneva observed a last glance from her mother, she grabbed Alice's hand. They ducked around the side of the church house and quickly approached the game of marbles in progress.

Kendall raised his head. He grinned. "Wanta play?" he asked Geneva.

"Sure."

"Oh, no, you don't!" Alice grabbed Geneva's arm.

"But I ain't gonna," Geneva continued. "An' you better not be playing with my blue taw er my other shooter, hear?"

"I will if I want to." Kendall grinned a knowing grin. "But I won't."

The girls watched the game as long as they dared, then went back to the women and children out front. Geneva picked out a girl from the Ridge group and plied her with questions about life up there. She caught a glance from Mattie. "I want you to meet my mamma," she said. "An' I'd like to meet yers. What's yer name again?"

"Shirley Barnsworth. I think I see my mamma talking to yer mamma now. Ain't she the one with the crochet book?"

"Yeah. She likes to do that stuff an' I guess yer mamma does, too. Do you?"

"Oh, it's all right. I have a hard time with it, but I reckon I'll learn. Mamma thinks I should."

Geneva laughed. "Hey, Alice. Shirley here's mamma thinks she ought to do that stupid crochet, too."

"I kind of like it," Alice said. "I guess because I think it's really easy."

"Come on, then. Let's go meet my mamma and get it over with."

Late in the afternoon folks began piling their stuff up to go home. "Chore time," the men called to the children. Women put leftovers and empty dishes and pans in their baskets. Kendall signaled to Geneva, and the two of them moved among the confusion to a quiet spot.

"Guess what, Gen? Those guys were telling me they is having a big championship marble tournament up at Toller's Store. You know that place, next to Eadsville Chapel? I'm gonna go. Want to come with me?"

"Think they'll let girls play?"

"I don't see why not. Course, you gotta get yer stash a marbles back an' that might be hard to do."

"I'll get 'em back, you just see if I don't, but I'm not sure I can get away from you-know-who."

Kendall grinned. "I reckon you'll think of something, if you want to go bad enough."

"What about Alice?"

"She might want to watch but she knows she ain't good enough to play in no tournament. You are, though."

"Yeah, I know. Can't think now how I'll do it, but if there's marbles to play I'll be there to play 'em."

"Gen-e-va! Geneva!" Mattie's voice rose above the crowd.

"Coming, Mamma. I'll talk to you later, Kendall. I'll go. I just have to plan how I can get there without Mamma knowing."

Kendall grinned. "You'll think of something fer sure."

"Reckon I didn't need my mules," Dade said to a Ridge man whose truck had been parked beside the wagon. "I brung 'em in case somebody got stuck in the soft road across the bottom."

"This week of fine weather 'bout dried up the tide. You can never tell about that road. I seen yer girl talking to mine. You bring her an the missus on up an let's visit some more."

Geneva realized Dade was talking to Shirley Barnsworth's daddy when Shirley piled into the back of the truck. "Geneva, y'all come up like Daddy says," Shirley said, "an' you an' me'll do that crochet." Shirley winked.

"Why, I'd like that, Shirley." Geneva smiled. Maybe this new friend wouldn't be so bad. Mattie came to the wagon. "Mamma, Shirley's invited me up to her house one of these days. Ain't that nice?"

Mattie nodded. "An' you come down, Shirley, when you can," Mamma added. "Bring yer mamma, too."

Mr. Barnsworth slowly backed the truck through

the thinning crowd. Geneva waved at her new friend and turned to Mattie.

"I'll see to Granny getting home, Mamma, an' then I'll sit with her a minute. I'll be back before milking time."

"That'll be fine, child. Mind, now, just because you been by Alice all afternoon don't mean yer to stop there this evening. Yer punishment ain't up till tomorrow."

"All right, Mamma. I'll come directly home."

Geneva found Granny still talking with Haw kin. "Can I carry anything home fer you, Granny?"

"That'd be mighty nice. Folks give me a heap a leftovers. They must think I can't cook ner bake no more. Look at all that cake. You'n me'll have us one more piece in a minute." Geneva thought Granny walked mighty sprightly along the short path to her house.

"Ain't you tired, Granny?"

"Not a bit. Seems like cutting that mop a hair off give me new strength. I done always heard growing hair takes it out of a body, an' now I believes it."

Geneva and Granny sat with their pieces of cake on Granny's porch and watched and waved as the last of the crowd went home through the river bottom or up the ridge.

"It's been a fine day," Granny said. "Reckon that's my last time to eat on the ground under them trees." Her voice was surprisingly cheerful.

"But, Granny, you said you was feeling better."

"I am, but that don't make no difference. I done told you I won't be here fer the last tide, an' I won't."

"What if there comes up a big rain in, say . . . June, or July. We might could get another tide. Sometimes that old river does it twice—or more—in one year."

"An' sometimes we have years on end that don't have no tide at all." Granny nodded.

"True. So how can you say fer certain you'll be gone before the next one?"

"A body knows those things when they's as old an as wise as this old woman is." Granny grinned. "When a body's this old, she can claim to be as wise as she wants."

Geneva laughed.

"I hear you been playing marbles," Granny said.

Geneva laughed again. "Yeah, an' Mamma gave Kendall all my marbles. Ones I won fair an' square. An' she gave him my blue taw."

"Likely she thinks yer too big to be playing with the boys."

"I reckon. Do you think I'm too big?"

"I don't know. Most folks would, I 'spect, an' maybe I used to would think that, but seems like I been studying things different lately."

"What do you mean?"

"Seems like times are a-changing an we better be a-changing with them."

"How so?"

"Folks seems to know they don't have to work from dawn to dark forever to get ahead. Seems like they're easier in earning a living than they used to be. Seems like that Bart might have him a good idear to go to town an' open him up a store. Don't reckon I know rightly what he aims to sell, but that'll be a sight easier surely than farming. Now, take yer daddy, he don't know nothing else. Yer daddy better be a farmer all his life 'cause when he quits he will surely die."

Granny paused. Geneva said nothing.

"That's why I'm a-telling you to mind that you don't put off to heaven fer all yer treasures. You get you a few of them now, like making you a teacher."

"I'd like that," Geneva agreed. "But Daddy won't talk about me going to high school an' I surely have to do that before I can make a teacher."

"I think you'll find you a way. Now, here's a switch. Yer mamma'd be mighty proud if you was to make one."

"I know. She thinks I ought to go to high school. Don't reckon I can, though. Her rheumatism is getting bad an' she needs me."

"Always the way it is with women. Somebody needs them, so they don't do what they want."

"I know." Geneva sighed.

"Sun's dancing on Tillet's Mountain." Granny nodded toward the west. "Another day's done."

"Reckon I'd better be going."

"Reckon you better. You come back soon. We's great ones fer talking, ain't we?"

"Seems like yer mighty easy to talk to fer such an old woman," Geneva teased.

"An' you, fer such a young one," Granny responded.

Geneva bent to hug the small woman. "I love you," she said.

"Pshaw, child. Don't be silly now."

Geneva hurried along the low road. Yes, she did love Granny. And her mamma, and her daddy. And certainly Alice and all of them. Yet these were a few short words that were never spoken. I wonder what is so hard about saying those words? Maybe they're words meant for only one special person . . . like Granny fer Granddaddy.

Geneva supposed they might have been said between her mamma and her daddy, once upon a time. But she could more easily imagine Ina and Bart saying them than her mamma and daddy—especially her mamma.

6

JAMES AND KENDALL, ALREADY ALONG the path through the pasture field, led the way. The three girls dawdled on the road, a long space of silence between them.

Betty Lou tugged at Geneva's hand. "Them people up there on Toller's Ridge didn't know you wasn't a boy, did they? Was I the only one who thought you was?"

"I reckon," Geneva growled.

"And why are you so pouty, Miss George-Geneva? It was all yer idea to dress up in them clothes an' try to fool everybody with being a boy. Playing marbles an' all." Alice demanded.

"I ain't pouty," Geneva grumbled.

"You are too, an' you don't need to be taking yer mads out on us." Alice said. "Come on, Betty Lou. Let's leave this pouty thing alone." The two girls ducked under the barbed-wire fence. Geneva, sliding under, got the seat of her overalls caught on a barb.

"You could at least unhook me," she called to

Alice, who was dragging Betty Lou through the cow pasture. The girls came back.

"Where you gonna change yer clothes? Or are you gonna let yer mamma see the way you went up to Toller's Ridge?" Alice said.

"Drat it all, Alice, unhook my rear end and I'll go to Granny's. She'll help me out an' she won't nag me about it neither."

Alice pulled the wire out of the overalls and the girls stood up, facing each other across the fence. "I'll come with you," Betty Lou said.

"No, you won't." Alice pulled the child away. "I told Mamma we'd come right home an' it's already late."

Geneva frowned at the moving backs.

"Y'all stop by my house an' tell Mamma I went to check on Granny an I'll be home directly," she called.

Alice waved an agreement.

Geneva dropped her eyes to the road and stomped away. The whole thing had been a stupid idea.

At Granny's, the old woman waited on the porch. "Where you been in them rags?"

"Me an the others been up on Toller's Ridge."

"You wearing overalls up there? Does yer mamma know? What you doing up there made you think you needed pants? Times may be a-changing, child, but you went off in *pants*?"

Geneva pulled two stockings filled with marbles out of the back pockets of the baggy overalls. She sat down. "We went up to a tournament. Kendall heard about it from some kids that was at the church's big spread. I figured them boys up there wouldn't let no girl play, 'specially when they could see she'd beat them. An' I was right. I didn't get to play. Then me an' Alice an' Betty Lou tried to fool Shirley Barnsworth that I was a boy come to call, an she wasn't fooled, neither. Matter of fact, she sat right there an' told Alice I was boy-crazy an' had done *everything* with a boy. She turned the tables on me.

"Then Kendall got to the very last game where he could a won. Them people up there was so put out that a boy from somewheres else might win Kendall decided he'd better not an' he threw the game to the other feller.

"We done all that an' went all that way fer nothing." Geneva threw her arms across her chest and frowned.

Granny chuckled. "Well, honey, you can go up there next time an' you can win. You know you can, so don't blather on about it now."

Geneva pushed the sacks of marbles with one finger. "Don't reckon I ever will. I reckon Mamma has won me. I'll not play no more. I might as well give them marbles to Kendall an' forget it. Yeah. Reckon that's what I'll do." She pulled her large straw hat off and turned it over in her lap. She poured the marbles into the crown and rifled her fingers through the bright bits of glass. She held an almost tigereye up to the light. " 'Cept this one. I'll give this one to James. An' here's my shooter, I'll give that to Betty Lou. I'll give my best taw to little Tad. He'll be a-playing marbles one of these days—I reckon town kids plays marbles."

She tucked the special ones into the top pocket of her overalls. "Now what am I supposed to do with the rest a them things?" She raised her eyes to Granny.

"You keep them, child. They's treasures on earth. You need you a blue jar."

Geneva began putting the marbles back in the stockings. "I don't want these stupid things no more. Give me a blue jar an' you can have these stupid earthly treasures."

Granny rose stiffly. "Wait a minute, Granny. I'll go down the cellar an' fetch one fer you."

"You stay out a my cellar. I'll take care of my own blue jars an' my own treasures. I'd be proud to have them pretty things, though. We didn't have such when I was coming up—course, we had marbles, but they wasn't pretty glass." She disappeared only to return in a moment. "I found me a jar in the kitchen.

"Now you pour them fellers in there slowlike er the jar'll break. That jar is old—it's a treasure by its self."

"You do it, then," Geneva handed the marbles to Granny. The old woman filled the jar one marble at a time, then held the full jar up to the light.

"My, Geneva, just you look how pretty them are."

"Yeah, but I don't care about them no more. What you gonna do with them, Granny? Go up to Toller's Ridge an' play you a tournament?"

"I'll set them in the kitchen winder, that one where the sun gets in first thing of a morning. Yes, sir, I'll have me something pretty to see."

Geneva rose. "I gotta go home."

"You going in them pants?"

The girl looked at her legs. "Naw, I just forgot I come here to put my dress on. Say, Granny, can I leave them here until tomorrow?"

Geneva tramped along the high road. She kicked at a loose rock and watched it careen down the hill. "Why are you so pouty?" Alice had asked.

I got plenty of reason, she thought. This day that should have been such fun was—had been—turned out—silly. Geneva sighed. It was all Mamma's fault. *She* ruined marble playing fer me. *She* won this game today. I might as well become that lady *she's* decided I'll be.

7

DADE STOPPED IN THE KITCHEN FOR A bite to eat. He sat at the table with his cold coffee and leftover biscuit. The April sun streamed through the window and a fly buzzed against the curtain. "Geneva, fix me a lunch. I'm going to the low field and I'll not be back fer dinner till I get done. You pick up the mail, child, an' help yer mamma."

"Yes, Daddy."

Mattie smiled. She likes this news, Geneva thought. She may think she's winning the battle to make me a lady 'stead of a boy child.

Geneva filled a jar with water. She tied an egg sandwich and a fried pie into a cloth. She added a handful of dried apples. She gave Dade the lunch, took a few coins and a short list from Mattie, and gladly left the house.

She pulled in a long breath of freshness and spring. The circle of mountains against the high blue sky was so pretty it almost hurt to look. Redbud trees, full in bloom, and the dogwood trees, veiled in white, pulled her eyes along to see every spot of beauty. She'd stop by Alice's, maybe they could go together.

Mattie opened the door. "Come right back, Geneva. Don't you be spending this pretty day fooling around."

"Yes, ma'am." But to herself she thought, if Alice can go, we'll cut across Castle Rock instead of going by the road. Geneva's eyes sought that rugged, brooding peak above and beyond Alice's. The cliff face, separated from the mountain behind it by years of erosion, stood out like a deserted castle. The children considered it a haunted place and had been warned of the dangers: shifting rocks and snakes.

Alice was in the garden. The baby Sissy sat in a tub nearby. She chewed on a thread spool tied around her neck and laughed at the world. Alice amused the baby with noises and singing as she thinned a long row of carrots. "Howdy," Geneva called. "Daddy's sent me to the store. Can you come along?"

Alice looked at the baby. "Don't reckon I can. Mamma's got me watching Sissy an' pulling out the extra carrots. I wish I could. Reckon Betty Lou could go, if yer wanting company."

"Naw. She's all right, but if I can't have you for company, I'd rather go alone. Tell yer mamma I'll fetch yer mail as usual."

Geneva strolled on. She turned for one last wave to Alice. I wonder how many times I can get her to wave, she thought, and turned again. Alice waved. Another few steps, another wave. Several more turns and waves brought Geneva to the giggles. That girl, she thought, she ain't pulling carrots, she's watching me, else she knows my tricks.

At the fork in the path she almost wished she were turning down to Granny's. Well, she'd likely get to Granny's before supper. Lately, it had been her job to check on the old woman. Geneva didn't mind. It seemed Granny had a new line of chatter every day—plans for what-to-do-after-I-die, stories from the past that came to mind and had to be told, or instructions about this herb or that one to be used for medicines.

Geneva was a ready listener and always found a few minutes before bed to write things down. She didn't know why she would want all those old stories, but they seemed important.

Without Alice, it wouldn't be any fun to cross over Castle Rock, and anyway, they knew not to go that way alone. Not that *that* would make any difference to Geneva. If she decided she wanted to cross Castle Rock she would cross Castle Rock alone or with another. I know, she thought, I'll take the short path and call the doodlebugs. It's such a fine day, they ought to be answering.

Along the trail she came to a less-traveled path. She skirted through a piny section of woods toward an outcrop of rock below the Castle. I reckon that big rock fell off the cliff, she thought for the thousandth time. She always meant to ask her teacher or someone about the huge boulders lying here and there among the trees, but away from them she forgot. Geneva found the rock she sought. The doodlebug place was a stretch of sand in front of a large boulder that lay in a small clearing among the trees. Last autumn's leaves covered the doodlebug sand and rested in the crevasses of the big rock. Smaller boulders lay scattered in the clearing.

Geneva found a tree branch to use as a broom. First she swept off the top of the boulder, then she swept off the sand below.

Feeling a little silly, she stood on top of the boulder to begin the doodlebug call. *Maybe Mamma's right. Maybe I am too old fer this foolishness. Still and all, I'll do it this one last time,* she decided.

"Doodlebug, doodlebug, come up from the sand. Bring yer family and the doodlebug band."

Geneva scrambled off the boulder. She peered at the swept-off place beneath. The sand lay still and soft in the morning sun. *It must be too cold fer doodlebugs,* she thought. *I'll come by here on my way home. Sun'll warm them doodlebugs by then.*

Every spring, just after the dogwood trees set flowers, a cold night would warn the river-bottom folks that winter might not yet be over. When the night of dogwood winter had passed, it was time to call the doodlebugs. *Likely Mamma done this when she was a girl,* Geneva thought, *and I know Granny did. Why, tomorrow er the next day Granny will be sure to ask me if I saw any doodlebug families. I'll have to tell her they ain't out . . . winter ain't over yet.*

"But it feels over," Geneva shouted, and flung her

arms overhead in a wild "Yahooo-ey!" She broke into a smart trot and bolted along the path and down.

Back on the road to the post office she ran and ran until, breathless and giggling, she came to the last turn before the store. She stopped, drew in a long breath, and smoothed her hair. Get yourself together, she thought. Them fellers in the store'll think yer granny-daft if you go in there like this.

Cousin Sheffield Haw greeted her: "Howdy, Miss Geneva. How's folks up to Dade's an' Bart's? How's Granny Haw?" Geneva knew her granny was not Sheffield's grandmother but everyone for miles around called the old woman Granny. Geneva reckoned Sheffield was a second cousin, maybe a third. It didn't matter. She knew half the people in the bottom were kin.

The store was full of folks who had to be greeted. Geneva exchanged the well-worn words: "Y'all right?" "Right as rain." "How's Miss So-and-so?" "Well, she's poorly." "How's yer daddy?" "Working hard." She knew every line, as did the others. She also felt their real concern for kin and neighbors alike.

Sheffield's hands and eyes moved among the mail in the sack. He handed a letter to this one, another to someone else, until the crowd gradually thinned. The twice-a-week delivery from town was excitement river-bottom folks loved.

"Here's yer pile," he said to Geneva. "One a them

gov-ment letters fer Granny. What y'all gonna do with that old woman?"

Geneva sighed. After the first two or three government letters Granny had refused to look at any more. "I know what them men want," Granny'd say. "No use reading another one a them things. If they think a little old letter's gonna drive me off this place, they better get them something else to think about."

Sheffield smiled at the girl. "Don't you be worrying none about it, child. Yer Daddy an' Bart'll take care of it."

Geneva sighed again. She'd go home and show Daddy the letter, but she knew it wouldn't be the men who'd solve the problem of Granny. It'd be up to her. Geneva took the pile of mail, said her good-byes, and went to the door. "Oh yes." She turned. "There's a few things Mamma wants."

Geneva could see Granny on the porch long before she thought Granny saw her. Halfway down the hill she paused. Was that a gun across Granny's knee?

"Heard you coming a mile off," Granny said, a dark frown above her normally interested black eyes. "You wouldn't make a very good Indian even with that Cher-o-kee blood in ya." Geneva had always heard that one of her ancestors was a Cherokee, and she believed that accounted for Granny's dark eyes.

She found a place to sit on the steps. "What'er you doing with Granddaddy's shotgun?"

"Ain't his no more. Mine. Cleaning it, what's it look like?"

"You sound a might put-out today, Granny. What's the matter?"

"I had me a dream last night. I dreamed them gov-ment men was a-carrying me off. There was four a them. Big brutes, an' they had me by the arms an' legs an' they was letting my bottom bounce along the road. They was laughing at my screaming an' yer daddy an' that Bart just stood there a-wringing their hands like some old fudd-dudd-don't-do-nothing-but-cry. So I'm expecting them to come on me, today er tomorrow, one."

"I don't know, Granny, but here's a letter from them now."

"Don't want it."

"Daddy said you was to read it."

"Ain't gonna."

"Then he said I was to read it to you."

"Don't bother. Won't listen. Ain't mine."

"But Granny, it *is* yours. Lookey here. It says, 'Mrs. Augusta Haw' right here on the front."

"Name ain't Augusta." She let her lip curl around the *Augusta*. "I'm Gussie. Get outta here an' take that fool thing with you."

Geneva got up to leave.

"Oh, child, I don't mean for you to go *now*. Take it home with you when you go. I think there's 'Augusta Haws' lives yonder, over there on the other side a that river."

Geneva sat down. "I was teasing you, Granny, same as you was teasing me. This here's yer letter an' you knows it."

"No, 'tain't."

"My daddy said if you wouldn't read it I was to read it to you. I'm gonna do that, Granny. Either you take it in yer hand an' read it or I'm gonna open it up right now an' read it to you."

"What you do don't make no difference to me. You ain't always done what yer daddy told you. How come yer doing it now?"

Geneva laughed. "You win." She put the letter in her pocket. She would find a place to leave it—on the kitchen table, or the bed, or someplace. That was about all she could do between Granny and the letter.

At supper, a grim and morose Dade quietly chewed his food. Mattie, equally silent and withdrawn, motioned directions to Geneva, who tried to think of something—anything—for them to talk about.

"I tried to see the doodlebugs on the way to Granny's," she said. Mamma smiled a half smile. Dade gave no indication that he heard.

"Can I find some fiddlehead ferns for salat greens tomorrow, Mamma? I think there's still some in the high places. We ain't had but a mess er two, and wouldn't we like them again before summer?"

"That would be fine, child."

"Did yer granny read the letter?" Dade asked.

"No."

"Did you read it to her?"

"No. She, uh, she had her shotgun on her lap an' I thought I'd best not make her mad. You know how daft she talks sometimes."

"What was that old woman doing with the shotgun?" Mattie cut in sharply.

"She said she was cleaning it. She said she was expecting the gov-ment men to come up on her an she wanted to be ready fer them."

"Oh, Lordy," Dade muttered. "It'd be just like her to shoot one a them. Then we'd really be in hot water with the gov-ment."

"She said she'd been having this dream about them gov-ment men—four of them, carrying her off an' bumping her backside on the ground." Geneva hoped for the laugh she got.

"That's about what'll happen," Dade agreed. His smile faded. "Geneva, I've about got my hands full here what with the crops an' moving an' all, an yer mamma's 'bout as busy as me. I'll guess we'll leave it up to you to see about Granny every day. Don't reckon Ina has time, what with her babies, an' that Bart can't think a nothing but his store in town. Looks like it's all up to you, gal."

"That'll be all right, Daddy. I like Granny. I get along good with her."

"Seems like you can do more with her than anyone," Mattie agreed. "Yer daddy an all of us is praying an' studying what to do about that woman. You can help by telling her what's gonna happen when the tide comes up on us, and by telling her *before* the tide gets here."

Geneva had hardly tucked her long braids under a straw hat and hooked a hoe under the first weed in the garden before she heard the unusual sound of a truck away off on the high road.

It wasn't any truck sound she recognized. The question of who had hardly entered her mind before the shiny yellow of a government truck came into view.

Oh, oh, she thought. I'll bet they're here about Granny. Geneva dropped the hoe and raced to the house. Daddy was far out of sight in the low field. Mamma was nowhere to be seen. Geneva slammed out of the house. She caught a glimpse of her mamma up in the little patch of ground where her long-dead baby brother lay. Geneva decided that going to Granny's couldn't wait for permission or announcement.

Geneva ran the low road. If she could make good time, she'd get there before the men were shot, for she had a picture in her mind of Granny, the shotgun in hand, meeting them.

Sure enough, one man was outside the truck and the other waited in the cab, while Granny stood defiant on the porch, gun across her arm but at the ready.

"Granny! You put that gun down!" Geneva called.

The man turned, removed his hat, and scratched his head. "You know this old lady?"

Geneva caught her breath. "She's my grandmother. Granny! You put that gun down, hear?"

"They come to carry me off just like in that dream."

"No, lady, we aren't gonna carry you off. We're here to help you understand you gotta leave pretty soon. Woodcutters'll come this summer an' all buildings have to be torn down to eighteen inches by January first, 1949. Now, you haven't accepted the money fer this place an' you haven't answered yer letters an' we haven't heard from yer kin, so we come to find out what we can do to help you." The man's voice was patient, with a sharp edge. "We'll be moving them graves over there an' we gotta know where you want yers to go. They can go to Clinton County, er Russell County, er up town, or wherever, but they gotta be moved soon."

"You ain't taking my man nowhere. I ain't going nowhere, 'cept to die, an I'm gonna be buried next to him."

"What's she talking about?" The man turned to Geneva.

"She says she's gonna die before the tide comes, an' she'll be buried right here by Granddaddy," Geneva reported.

"She don't look like she's about to die. Is there something wrong with her?"

"I don't know. I reckon she's dying of a broken heart."

He snorted a short laugh. "People don't die of broken hearts. But if she's gonna die, we could give her a little more time. Does she understand about the woodcutters? An' getting buildings moved er taken down? And us moving graves?"

"I don't know. We've tried to talk to her, but mostly she won't listen. She just says she's gonna die an that'll end that."

"Yer just a kid," he said. "I better go talk to somebody else. Who's her people?"

"My daddy's along the low road in the field over yonder. My mamma's to home—that way, see that big white house? You can go low road er high road through the woods. My Uncle Bart an' Aunt Ina er in town today seeing about their place. They'll be moving soon, I reckon, when the crops er done."

The man turned to Granny. "Now, Miz Haw, we're gonna talk to yer son. We'll be back. Crew's coming in here soon to move those graves. We know where most of them are to go, but we don't know about some of the Haw graves. You decide, an' we'll be back in a day er two."

"Git. Git off my yard. Don't come back no more. I'll fill yer backside with buckshot, that's what I'll do. I'll shoot out them tires. That'll slow you down some."

The man shrugged his shoulders and climbed into the truck. Geneva watched him back out and turn to go down the low road to where Dade was working in the field. She took the gun from Granny. "Sit down. Yer trembling like a leaf."

Granny sat. "I done drove them off. There they go, scared to death."

"Granny, you didn't drive them off. They're going down to see Daddy. You've got to move pretty soon, Granny, but he said you could stay a might longer. They're coming back in a few days to move the graves in the burying ground. You gotta decide where they're to take Granddaddy an' the others."

"They ain't taking my man anywheres. I done promised him he'd be right there where I could meet him so we can spend eternity in our heavenly home. He ain't going no place, an' neither am I."

"Then what er you gonna do when they come back with a big crew a men, all with shovels ready to dig up them graves?"

"I'll shoot 'em. Every last durned one a them."

"Then they will take you off to jail. Now, think a that. Jail."

"Oh."

"You better decide. Where're they to take Granddaddy Haw an them other Haws?"

"Don't care where they take them others. I promised my man . . ."

"Do you want him to stay here? Deep under the tide? They say the lake'll be six–seven hundred feet deep here. That's a lot a water. You won't be able to visit, er bring flowers, er be buried here neither."

"Oh."

"What do you want to do, Granny?"

"I'm gonna die soon. Surely the Lord will call me in time to be buried by his side."

"I don't think they'd let you be buried there, an' if they did, you'd be moved soon, too. Those men er coming in a *few days*. You gonna die in a few days? How do you feel? Are you hurting someplace bad enough to die?"

"No, I ain't hurting at all. All but my heart. It hurts bad."

"The kind of hurting that is dying?"

"No. Not that kind a hurting. Hurt an' broke, just like you told that feller. Dying of a broken heart takes a long time."

"Well. I understand. So we've got a while to take care of yer dying. In the meantime, we gotta do something about Granddaddy."

Geneva paused. Granny frowned.

"I'm studying this, gal."

"The choice is, where do you want him moved er what can we do to keep him from being moved if you want him here under six hundred feet of water. Maybe we ought to ask Daddy er Uncle Bart."

"No. It ain't their decision."

"Yes, it is. He was their daddy."

"But he was my man. I'll decide."

"Okay. You decide."

"I promised him."

"Yes. But if those men find his grave marker, they'll take him. It's the law."

"If they don't find his grave marker, he'll stay?"

"I reckon."

"Then we gotta move that body-rock."

"Granny! That headstone'd weigh a *ton.*"

Granny grinned. "We got us a job to do, girl."

"Now?"

"Now.

"Granny! Yer old an' I'm young. How're we gonna move that body-rock?"

"I moved plenty a stones in my day. Why, lookey them stones in the church-house foundation. Who do you think fetched all them stones? Women an' children, that's who. You get you a good lever, an' a stump fer a fulcrum, an' it ain't hard. Let's go 'round back an' see what's in my woodpile."

Geneva rolled her eyes heavenward. What would this woman think of next?

"Grab you this grub hoe," Granny handed Geneva a rusty picklike tool. Granny studied the woodpile and then pointed to a thick, dry sapling. "Reckon that'll do. Load that up."

"Into what?"

"That there wheelbarrow." Granny pointed to a thin metal wheel with part of a wooden box attached to some handles. "Yer granddaddy made that thing an' we'll use it to keep our promise."

The old woman cocked an eye at her chopping block. "That thing's too big," Geneva said.

"Reckon it is. Here, put in that rock. That split a wood, an' that big stick. There's another rock. Ain't no good ones in the burying ground. Let's go."

Geneva hefted the handles of the barrow. She groaned. Granny laughed. "'Tain't funny," Geneva muttered.

"Lucky it's most down hill," Granny answered. "Here, I'll carry the grub hoe an that lever-sapling."

Geneva pushed against the barrow. "Atta-girl," Granny encouraged. The wheel began a slow turn. Geneva wobbled and coaxed the lumbering thing around the house, through the gate, and along the path to the church house. Once started, the barrow moved easily. In a moment the girl and the barrow careened past the steps, around the corner of the building, and on to the burying ground. Granny, laughing with it all, careened along behind.

"It ain't funny," Geneva shouted.

" 'Tis so. Anyway, better to laugh than cry about it." Granny hooted. Geneva headed for the corner where the deed was to be done, tipped the barrow, and let the rocks and chunks of wood spill out among the jars of faded flowers and wreaths of dried greens left from Decoration Day. Granny stopped in front of Granddaddy Haw's resting place. Her shoulders drooped and Geneva stepped back to allow the old woman a moment of quiet.

The old shoulders pulled in a long sigh. "Come on, child. Let's do it." She picked up the grub hoe and savagely chipped away at the dirt and grass growing around the base of the monument.

Geneva looked at the square gray rock, three feet tall and a foot wide. Granddaddy Haw's name and dates had been chipped into the stone and polished to a high shine. Geneva read again the words beneath his name: "Waiting in his heavenly home." Granny's name was below his but no dates were yet inscribed for her. The old woman's puffing caught Geneva's attention. "Here, Granny, let me do that. You figure out what'll happen once we get it loose."

Granny walked through the small burying ground. A white-painted fence formed three sides and the fourth was open to the woods beyond and below. Geneva paused to wipe the sweat on her shirttail. She saw Granny scan the distance from the headstone to the woods beyond.

"We'll roll 'er over this way an' tumble er down over the edge. Them gov-ment men'll not look into the trees."

Geneva bent to the digging. "Get more out of the back," Granny advised, "so's it's tippylike."

A few more strokes and the top half of the monument fell over.

"This thing's in two parts!" Geneva exclaimed.

"That'll make it a sight easier. We'll just get rid of the top an' cover over the rest. We'll have to hide our tracks. Ain't nobody's business what we been doing here."

Geneva caught her breath. "Hot, ain't it? Feels right muggy. Reckon it'll rain this evening?"

Granny glanced at the sky. "Reckon. Be fine if it does. Help cover our tracks."

"Now what?" Geneva asked.

"We'll use this small rock we fetched along an' we'll put our lever-sapling where you dug out an' you'll see how easy it is to roll that body-rock over the hill."

Geneva and Granny bore down on the lever-sapling. The rock rolled sideways. "See?" Granny chortled. "Now that wasn't hard, was it?"

Geneva caught her breath. "Not harder than common. Reckon we can do it."

Once beyond the clearing the rock was easier to roll. It jagged down the incline and with one last push reeled out of sight. Granny dusted off her hands. "That did it. Now I can keep half a the promise. I'll have to study on the other half."

"Study on what?"

"About my burying here by him."

"Granny, you got to die first."

"I'm working on it, gal."

"Come on, we gotta hide our tracks." Geneva put her arm across the old shoulders. Granny is as thin an' fragile as the English bone-china cup so prized by my mamma, she thought. I bet I could see right through her, just like I see my finger behind the cup rim.

Back at Granny's, Geneva paused to drink cold, fresh water from the spring. "I been gone a long time," she said. "Reckon Mamma'll think I'm off playing with Alice. What'm I gonna tell her, Granny?"

"Tell her nothing."

"You don't tell her nothing," Geneva answered. "With my mamma you tell her something."

Granny cackled in appreciation. "Well, tell her you come down to save them gov-ment men from getting shot."

"Yeah. That's a good one. Mamma'll have heard about it from Daddy, I reckon, time I get on home. What'll I tell Daddy about what we done?"

"Tell him nothing. If—when—he comes down to see about his daddy's grave, I'll tell him. But likely he won't come till after them grave robbers is here."

"They ain't grave robbers, Granny. They're grave movers."

"'Mounts to the same thing. You tell yer daddy you talked me into signing them papers so them other Haws can be moved. That'll explain how long you been here. Fetch me that envelope from the mantle, child. I'll sign that thing now. Get in that little drawer under the clock an' bring me that in-del-ible pencil. Come on now, 'fore I change my mind."

Geneva watched the thin, wrinkled fingers form

the letters of Granny's name. "I don't write pretty like yer granddaddy done," she muttered.

The brightness of the day dimmed. They heard the low rumble of thunder against Tillet's Mountain to the west. "Git on home," Granny commanded. "An if it come to light-ning, don't you be stopping under no tree."

On the porch, Geneva stopped to hug the old woman. "Git!" Granny said. "Git on out a here now."

Geneva raced out of the yard. She knew she had plenty of time before the storm broke. She also understood that Granny's gruffness was her way of closing this part of her life.

8

GENEVA AND ALICE, BLISSFULLY ALONE, drowsed in the hot, muggy, summer Sunday afternoon. To get away from the younger children, they had walked up behind Geneva's house to the little patch garden where Geneva's baby brother had been buried so long ago.

Geneva had related her adventure with Granny and the government men and the headstone moving. The girls had discussed Granny's motives, and Geneva had wondered, also, about this little place so carefully tended by her mamma.

A silence stretched between them. Geneva sighed.

"Yeah," Alice agreed. "A couple more weeks."

"How'd you know what I was going to say?"

" 'Cause I was thinking the same thing. School'll take up in a couple more weeks."

"Our last year at Hawford School."

"I won't even get a whole year. We'll be going to town in September."

"An' we'll be going to Toller's Ridge shortly after."

"It would be our last year, anyway," Alice comforted, "us being in eighth grade an' all."

"Reckon neither of us'll get to go to graduation. Daddy says I might as well quit school when we move, no use starting up there an' only go a month er two. Mamma says I'm to go the same as always."

"I wish you could come to town with us. I'm going to high school, an' work in Daddy's store. He says if Mamma needs help with the babies in town she can hire help easy. But he says she'll have it so good in town what with the electric an' a new stove like he's going to sell an' a new refrigerator an' everything she won't need no help of any kind."

"Daddy said I don't need no new clothes for school," Geneva complained. "He says it'd be a waste a money."

"I wonder why he don't want you to go to school?"

"He says I don't need no education to be a farmer's wife an' anyway, one of these days, the new farm on Toller's Ridge'll be mine an' I better find me a good man so's we can run it."

"But, Gen, yer only fourteen."

"Yeah. He says lots a girls get married at fifteen er sixteen, but I can wait till I'm bigger. Mamma says I don't need to get married at all. She says she agrees with Granny, I ought to make teacher. They wouldn't hire me as teacher if I was to be married."

"I think he has some other reason fer not wanting you to go to school," Alice observed.

"Me, too. I been studying on it. I'm gonna ask Granny what she thinks one of these days. When

Granny ain't talking daft, she's full of good ideas."

"She mostly talks daft to me," Alice said.

"I think a lot of it is playacting. She can keep people from prying into her business if she talks daft."

"We made our catalog orders," Alice said. "Daddy took it to town yesterday to the order store but Mamma says we will start school here in our old clothes an' save the new ones fer town school."

"I was looking at some of them dresses. Most of all, I was looking at how the girls in the pictures has their hair done."

"Yeah. I think Mamma'll let me get mine cut when we get to town. I think she's afraid what yer mamma'll say to her if I get it cut now."

"Hey! Remember the day we decided to cut our hair when we saw Granny with hers cut? Let's do it now."

"Now? Yer house an' mine er full of people talking. Where could we go?"

"Out in the barn? No. Behind . . . I know, let's tell them we're going down to see Granny. Then we can go up to Castle Rock an' do it up there."

"Let's each fill a poke . . ."

The girls jumped up. "Be careful," Geneva warned. "We don't want Betty Lou coming with us. I'll meet you at the secret place in five minutes."

"Here's the doodlebug rock," Geneva said. "I swept it off one day last spring. What'll happen to the doodlebugs when the tide comes up?"

"Let's call them one more time." Alice began to chant. "Doodlebug, doodlebug . . ."

"Alice, stop it. We're too big fer that silly stuff. We come here on serious business. Let's stop right here an' do it. What did you bring?"

Alice opened her poke. "I got them sharp scissors Mamma uses to cut the boys' hair. An' here's a comb. I brought some soap, too, so we can wash our hair after in the spring up yonder."

"I got a mirror an' my comb an' a picture from the catalog. Here's some towels. Who's gonna be first?"

"I'll go first," Alice said. "My mamma ain't gonna be near as mad as yers. Then if we don't like it, we don't have to cut yers at all." Alice started to undo her long braids.

"Wait a minute. Wait a minute. Granny cut her braid off braided. She winds it back up there sometimes," Geneva said.

"I know," Alice agreed. "Let's rebraid it in one braid. It'll be easier to cut an' better to save. I'd sorta like to save my hair, wouldn't you?"

"I brought a couple a paper sacks so we can."

The girls released their long braids, combed out the snarls, and rebraided the hair into one thick braid.

"I won't have it to make that funny mustache anymore," Geneva said. "Nobody'll laugh at me when the teacher's back is turned."

"An' you won't have it hanging in yer marble game a-bothering you," Alice said.

"An' when I wash it it'll dry real easy."

"No more one hundred brush strokes before bed," Alice crowed.

"No more snarls."

"An' cooler under yer hat."

"Yeah, cooler in winter, too."

"No more ribbons to wash an iron fer Sunday."

"We're too big for ribbons anyway. Only little girls wear ribbons."

"Here, sit down below me," Geneva directed. "I'll cut yers then you can cut mine." The scissors sighed into the first hairs. Alice squealed. "Hurry up. Before I change my mind." Geneva paused. "Cut! I ain't gonna change my mind."

"Me either." Geneva sawed the scissors through the thickness of hair above the braid. Bit by bit, fine strands of hair fluttered around Alice's face. She held the mirror and watched. "I love it. It looks *pretty*."

"It looks cute," Geneva declared, laying the long braid on the towel in Alice's lap, "but it don't look like the picture of the girl in the catalog."

"She's got hers curly," Alice observed. "I ain't got curl in mine. But look, cut this off some here by my ears. This part's too long."

Comb in hand, Geneva worked on Alice's now-shorn locks. In a few minutes Alice was satisfied. "I can hardly wait to wash it," she said.

"Let's do it now."

"No. Let's cut yers first."

"Okay. Do it."

Geneva sat on the rock. She spread the towel across her lap. "I want it to look like that girl." She pointed to a girl whose hair fringed her forehead above the eyebrows. "That'll look good on me."

"Here's the mirror." Alice held up the long braid. "You sure?"

"I'm sure."

The scissors bit into the thickness of hair. The girls fell silent as Geneva's hair hung in damp waves around her ears. "Hot, ain't it?" Alice asked, gritting her tongue against her teeth as the cut sawed through the thick and heavy braid.

"An' I sweat so, in my hair." Geneva said. "Er, I forgot, 'ladies don't sweat, they perspire,' " Geneva giggled. "Anyway, it'll sure be cooler with this stuff cut off."

"Why, lookey here," Alice said as she lay the braid in Geneva's lap, "yer hair's got a wave to it. Mine is straight."

"I thought I had a wave in my hair all along." Geneva said. "Now, get busy on that little fringe across my forehead but don't get it too short. I don't want it to stick up."

Geneva closed her eyes as wisps of hair drifted to her nose and lips. She sneezed. "Don't do that!" Alice commanded. "You made me take a big gouge out of this side. I'll have to comb some more over here. Hold still, now."

"Can't help it. Hair tickles."

"You can look now," Alice declared.

"Oh, it looks wonderful," Geneva exclaimed. "I love it. Just like Granny said, it makes you free."

"Put the braid in yer sack. Let's find the spring an' wash our hair an get it combed to look like the picture," Alice said.

Well, it's done, Geneva thought. The easy part is done. The hard part will be facing my mamma.

9

"I'M GONNA HAVE TO PART IT OVER HERE." Geneva surveyed the gouge in the fringe of hair Alice had cut. The girls had washed their hair in the spring and had helped each other comb it out into what they considered beautiful styles.

"It looks good that way," Alice agreed. "Is mine pretty?"

"Oh, yes! An don't it feel free?"

"The shadows er getting long, Gen. I reckon we got to be getting on home." Alice sighed.

"An' face the music," Geneva said.

"You scared?"

"Yes."

"Why don't we run down to Granny's first?" Alice suggested.

"We'll have to hurry, but let's go. She'll like it, even if Mamma won't. She's been telling me to do it right along."

Granny called from her place on the front porch. "I hear the two a you coming. Come on up here an' get you a chair."

"Howdy, Granny." Geneva tried to keep the smile off her face. "Alice'n me been busy today."

"I see." Granny nodded. "Turn around here, let me take a look at ya. Well, now, 'cept fer that place on Geneva it looks right fine. You done it, huh, you finally done like I told you."

"Oh, yes, Granny, an' it feels so *good*." Alice giggled.

"Scared to go home, ain't you, child?" Granny looked directly at Geneva. "You won't have no trouble, Alice. Likely yer mamma'll laugh about it next time I sees her. But that other one . . . won't be any laughing matter to her."

Geneva swallowed. She almost wished the hated braid still hung down her back. Almost. But not quite. "Got any ideas of something to tell her?" she asked.

Granny cackled. "Ain't like you not to have something to say. I reckon you'll weather the storm, same as you always done."

Geneva laughed. "Thanks, Granny, you always make me feel brave."

"We better go," Alice cut in. "It's getting late. We just wanted to show you, Granny."

"I told you to do it. If them women up there was smart, they'd get rid a their long old stuff, too. Hair holds you down. Keeps you busy an' don't let you do the things needs to be done. An' getting it cut off gives you strength to do the hard stuff of life. Gives you strength to get them treasures in blue jars."

Geneva and Alice looked at each other. Daft talk, their eyes said.

"Bye, Granny. I'll see you tomorrow," Geneva said. "I'll tell you about my whipping."

"You won't get no whipping," Granny called. "Yer too big. Yer mamma gived you yer last whipping the last whipping she gave you."

"I'm glad to hear that." Geneva laughed. The girls waved from the high road. They would hide their sacks of braided hair in their secret place. One thing to face at a time, they had decided.

One scream then a stony silence greeted Geneva as she met Mattie in the kitchen with the full bucket of milk ready for the separator. Geneva ducked out and headed back to the barn with the empty bucket. I'll stay away from her, she thought, an' let her get used to the idea.

As for Dade, he was silent at Geneva's arrival and had remained silent as the milking continued. Geneva decided he had not noticed the hair. It was likely he would not notice, so morose and withdrawn her father had become. Geneva felt sorry for him but only partially understood the depth of his sorrow. She knew his deep attachment to the river, land, and farm buildings he had worked among all his forty-eight years. That they would be gone forever still did not seem possible to her.

As Geneva squeezed the milk from the cow's warm

udder she felt a slight breeze drifting through the barn and appreciated how it touched her hair-free neck. Whatever happened, she would never be sorry she cut her hair.

Silence greeted father and daughter at supper. Mattie refused to look at Geneva, Dade chewed slowly but ate little. Geneva waited for the blow to fall. She wished it would. This silence was harder to bear than any of Mattie's punishments.

Geneva automatically finished her chores and quietly went up the steps to her room. She would not join her parents on the porch this evening and she doubted that her mother would call her for Bible-reading time. Silence such as this was a warning as to the depth of Mattie's anger.

Mattie was not in the kitchen when Geneva entered with the morning's milk. That's unusual, Geneva thought. She glanced out the window, thinking her mother might have gone to the privy. Instead, she caught a glimpse of her climbing the path to the baby's burying ground. Mamma ain't said nothing to me since I come home yesterday, Geneva thought, an now she's going up there again.

Separating the milk was usually Mamma's job but Geneva knew how to do it. Geneva poured the boiling water over the parts of the separator, assembled the circles and disks, put the filter in, and poured the milk into the top vat. She clamped on the handle and began to turn it slowly and rhythmically. She heard the milk begin its journey through the separator gears and saw the first line of warm milk drizzle out the milk spout. In a few more turns the separated cream flowed from the cream spout. Daddy came in with his last pail. "Where's yer mother?" Geneva pointed toward the little burying ground. "Finish up here, child. She's grieving."

For a baby dead ten years? Geneva asked herself.

She finished the separating, carried the milk and cream to the spring house, and poured it into the clean cans that were waiting there. She put these into the icy cold water, took a long drink, and returned to the kitchen.

Quickly finishing the breakfast, she called Dade from the back door and went to the front door to peer up the hill toward the quiet form of her mother. Geneva hesitated. She decided not to call, but went in to give her daddy and herself something to eat.

Dishes done, separater again washed and rinsed in boiling water, Geneva looked for another task. Mattie was still sitting, quiet and forlorn, on the hillside. Geneva became aware of a truck noise on the high road. What now?

Geneva and Dade arrived in the yard at the same moment. They saw a bright yellow government truck, followed by several other trucks, stir up the dust as they headed along the road.

"Grave movers," Dade muttered. He turned back

toward the barn. "You give them that paper yer granny an' I signed. It's there on the mantle under the clock. I don't need to see them."

"Ain't you going down to the church house with them to see about it? Ain't you gonna say nothing to Granny?"

"Don't need to. She had me to sign the papers. It's over an' done. You go meet them up on the road."

Geneva took the paper and handed it to the man in the first truck. He seemed to recognize her. "That old lady gonna shoot me, girl?" He grinned a little sheepishly. "You better come along an' protect me."

"She'll be all right. She had Daddy to sign the paper, too." Geneva turned away. Joking with govment men was not what she wanted to do this morning.

She went to the kitchen. She added wood to the cook fire and set some beans in a pot to cook. Likely they'd not be ready for dinner, but she couldn't think of anything else to do to help her mamma.

Taking her courage in hand, Geneva followed the path to where her mamma sat, statue quiet, dry eyed and bent. As Geneva approached it came to her that Mamma was old beyond her years.

Geneva found a place to sit. "Mamma, how did my baby brother come to die?" She had decided the direct approach would be best. If it made her mamma mad, so much the better. Anger was easier to cope with than this all-pervasive silence.

"He was borned dead."

"How come?"

"I don't know. We waited eight long years for you to be borned an' you was easy as falling off a log. Four years later, when I realized there was to be another baby, everything went wrong. He come early an' he didn't live hardly a minute. Yer daddy made him a little box that very night and brung him way up here so's he'd not be in the tide, for it was rising fast that very night. I didn't want him way off at the church house. I wanted him close by me. An' there never was another baby to be had. No more."

"Are you still grieving fer him, Mamma?"

"No, not exactly grieving. Sometimes I wonder what he would be like. He'd be ten years old. I expect he'd be like that James, another Haw. No, I ain't really grieving fer him."

"Are you mad 'cause I cut off my hair?"

"No, child. I guess I always knew you'd do it someday. I see the stubborn Haw streak in you. I guess I can't make you into something you ain't aiming to be." Mamma paused.

"I don't know what to do. Law says he's gotta be moved before the tide comes. I ain't told the govment his little box is here. Likely, that little box is gone, an all that's left is his little bones. I don't want them digging him up. Let him rest in peace."

"Mamma, Granddaddy Haw ain't gonna be moved. The baby could stay here, too. He an' Granddaddy could be together under the water."

"What do you mean? Granny signed the papers.

Yer granddaddy is going to Russell County same as them other Haws."

"No." Geneva shook her head. "Granny fixed it so he'd stay here."

"Is there something you should have told yer daddy er me?" Geneva was relieved to hear the old sharpness in Mattie's voice.

"Two, three weeks ago Granny decided she didn't want him moved. I helped her tumble the headstone over the hill into the woods. We covered over the grave with brush an' stuff so you can't even see it. She thinks them gov-ment men won't move it if they don't know it's there."

"I expect she's right. Likely she'll stand on it an' watch them an' talk daft an they'll leave her alone."

Geneva was pleased to see a slight smile on her mother's drawn face. "Does your daddy know this?"

"No. Granny said I wasn't to tell. She said if Daddy come down there looking fer it, she'd tell him, or Bart. But she said less knew it, better it'd be."

Geneva and Mattie gazed across the top of the farm, across the barnyard to the field where they could barely see Dade and the mules at work. "I don't think yer daddy'd care. He'd say whatever Granny wanted would be the way it should be. He told me whatever I wanted to do about them little baby bones would be the way it would be fer him, too. Yer daddy can be as stubborn as a mule, but sometimes he won't make a decision about nothing that ain't to do with farming."

"Except that I can't go to high school."

"That, too."

"Why, Mamma, why? Why won't he let me go?"

"I think you said it, child. He don't want to *let* you go."

"I'd only be going up town. I wouldn't be gone to—to—Africa, or something like that."

"He's losing everything, child. Everything. Yer going to high school is just one more thing he's to lose, only he can say 'no' to losing that."

"I might go anyway."

"I expect you might. It wouldn't surprise me none about you."

"Granny said I should make a teacher. I'd like to be a teacher."

"I'd like it fer you, too."

Geneva touched her mother's hand. "Mamma, it would be all right fer the little bones to lie right here. The gov-ment don't need to know everything about our lives."

"It's the law. I was raised to respect the law, an' the rules, an' God's will. That's what I been studying—if it is God's will er mine to move him er leave him here."

"Granny said she promised Granddaddy he'd never move from this land an' she'd meet him right here to spend eternity in their heavenly home. That's why she decided to fool the gov-ment men."

"Even if yer granny was to die tomorrow, they wouldn't let us bury her down there," Mattie said.

"I know that, but she thinks when they go away she'll get her own way about the burying. I'd say we'll need a lot a good luck to do it her way." Geneva added.

"I don't think Granny's ready to die. I think we'll have to move her off before the tide."

"That's something else we'll have to have a lot a luck to do." Geneva smiled.

Mattie gave a small chuckle. "You run on home, child. I'll study this thing a few more minutes. I'll be home directly."

Geneva bent to kiss her mamma's cheek. "I love you, Mamma," she whispered.

"Pshaw, child. Don't be silly, now . . ."

Geneva sped down the path. She'd work ever so hard at helping Mamma. She'd do like she was told, and she'd not be willful and go her own way. . . .

Pshaw, child, don't . . . the same words Granny had used at her offer of love words.

10

GENEVA CHECKED THE FIRE UNDER THE boiling beans, added a stick of wood, and went to take care of the hens and chickens. She longed to know what was happening at Granny's, what with the govment men down there in the churchyard. Half of her wanted to go see what the men were doing, but better sense told her it was no place for her.

A couple of hours before supper the trucks pulled up the high road in a trail of dust, but Geneva could see no difference in them from morning to afternoon. Perhaps they had not done what they set out to do.

"Mamma, can I run down to Granny's an' see if she's all right?"

Mattie had gone about her work quietly all day, but it wasn't the stony silence of yesterday. Geneva felt the weight lift from her shoulders. Her mamma's peacefulness left the girl feeling as free as the haircut did.

"That would be all right. I suppose you already done showed her yer haircut." It was more of a statement than a question, but Geneva felt the need to answer.

"Yes, ma'am."

"I suppose Alice done hers, too. I suppose you talked her into it."

"No, ma'am, I mean, yes; she cut hers, but I didn't talk her into it. She talked *me* into it."

"Don't you be blaming her. Yer the oldest here, even if it's only by a month or two. You know better, er you should know better. You been raised different."

Geneva wondered what her mamma meant, but she wisely decided it was not to be questioned. "Do you want me to take anything to her?"

"There's a pile of mending on the hall table. Reckon she'd like to do that. She's always asking fer something to keep her fingers busy."

"Didn't she ever learn that crochet?"

"No. She said she couldn't afford the thread. That's a laugh. She always was a quilt piecer, but she's lately quit that, too."

"She says she's ready to die," Geneva said.

"Reckon she is. She can't get it into her head she's got to move and I suppose dying would solve all her problems, at least I suppose she thinks it would."

"I'll be back shortly." Geneva said.

Alice was in the yard with the baby. She had tied Sissy into an old tire swung from a low branch. Sissy squealed with delight as the swing glided back and forth. Tad worked at the pushing, too, though Alice had to moderate his energy. "Howdy, what'd yer mamma say?" Geneva asked. Tad ran to catch her knees in a tackle and she landed on the ground with a thump. She turned the child over to tickle his back.

"Not much. Daddy didn't even notice. How about yer mamma?"

"Yesterday she was plumb silent mad, but today she's sorta resigned to it all."

"Do you wish it was back on yer head?"

"No, an' I reckon it'll grow if I've a mind to let it. How about you?"

Alice ran a hand across the back of her neck. "I felt kind of funny when I put my head on my pillow. Seemed like part a me was missing. You on your way to Granny's?"

"I'm wondering if she's all right after them gov-ment men was there."

Geneva found Granny's door tightly closed. In fact, it seemed to have been bolted from the inside. Geneva pecked at the door. "Granny? It's me. Granny, are you all right? Granny!"

"I heard you, child. Are you alone?"

"Of course. Who would be with me? Come on, Granny, unlock this door."

Granny let the door slowly open. Geneva stepped inside just as the old woman put the shotgun in the corner. "Granny, what'er you doing with that gun?"

"Them gov-ment men might be sneaking 'round back to get me. They're bent on moving me where I don't aim to go. They been here all day, messing around that church house planning on how to do in this old woman. I fooled 'em. I was here with my gun an' I was ready fer 'em to try something on me."

"Granny, don't you remember? You told Daddy to sign the papers so they could move the Haw graves. That's what they was doing today, not planning how to take you off."

"Is that a fact? All that carrying on at the church house was them doing that? I reckon we better get on over there an' see if they found my man." Granny picked up the gun again.

"Granny, you put that down. You don't need no gun. We're only going to the burying ground."

"Some a them fellers might be waiting on me. It'd be just like 'em, trickylike."

Geneva took the gun. She checked the breach and removed the cartridge. "Granny, you had this thing loaded!" she scolded.

"Sure I did. I was ready, I tell you. I filled me up a bunch a buckshot casings and they'd felt it plenty if'n I'd needed to shoot."

Geneva slipped the cartridge into her pocket. She let her eyes sweep the big room in a search for the rest of the shells Granny had prepared.

"Won't do you no good to take 'em. I'll just make me some more. I can take care of myself."

Geneva sighed.

"What'd you bring me, child? Something to eat? I ain't had me a chance to get a bite all day, what with them fellers prowling around an' needing watching."

"It's some mending Mamma thought you might want to do. Come on, let's go to the kitchen an' get yer fire going. We'll find you something to eat."

"What'd you come down here fer, child?"

"I came to see if you was all right an' what them men did in the burying ground."

"Was they moving graves today? You sure that's what they was doing?"

"I'm sure. Let's go see, then you can eat."

Geneva walked briskly through the house, across the porch and around to the church house. Granny slunk behind, peering this way and that into the late-afternoon shadows. "Come on, Granny. I ain't got all day. Mamma said I should come right home."

"Them fellers might still be around here," Granny cautioned.

"No, they went. They went in their trucks."

"You sure?"

"Sure. Come *on*."

Around the building the mess was unimaginable. In a place that had always been neat and well cared for, chaos now prevailed. Granny went first to the place she and Geneva had tried to protect. Granny cackled. "*Hhheee,* they didn't find my man! I fooled 'em. He's still here. But would you look at this mess? Folks er gonna be mighty perturbed about this."

"I guess it don't matter. It'll soon be under the tide." For the first time, the full realization of what this tide meant to all their lives came to Geneva and she understood some of her mamma's and daddy's sadness.

Granny flared; "It does so matter. This is the Lord's place an it behooves us to keep it fitting fer Him."

"Can we do anything about it?" Geneva asked.

"Reckon the elders need to come by with them mules an' ground-blades an' straighten the place up. Then it'd need new grass seed, but hot summer ain't a good time to plant grass. Well, we'll do it anyway."

"Granny, we don't need to go that far in fixing it up. I'll tell Daddy an' he can smooth all this dirt an' rocks out again."

"No, we'll need grass, too. Folks hereabouts'll be by to look where their people was; it'll help 'em think about the moving if'n the old place was like it's supposed to be. You go on home, child. Send yer daddy down here in the morning an' I'll tell him what to do. I'll study this thing a few more minutes. I'll go home directly."

Geneva turned to go. Where had she heard those words before? "I'll stay here a minute"—her mamma's words. Women's words. Taking-care-of-others' words.

Geneva took the high road home. In the filtered light of the woods the hard reality of the changes coming—too fast now—weakened Geneva's courage and she let herself cry. Great, wrenching sobs, first, then slow, sad tears.

Dade, Mattie and Geneva sat down to supper in the last glow of sunset. "Ought to light a lamp," Mattie said.

Dade grunted. It was neither an agreement nor an objection.

"No, Mamma, don't."

In the quiet evening the sleepy calls of the nesting birds, the lowering of the cattle and the rustlings of the fowl seeking shelter comforted the girl. She felt tears rise again. She studied her daddy's face. If he were not a man, she decided, he'd cry, too.

Mamma's tears seem all used up, Geneva observed. Reckon she cried it all out up there where the baby bones lie quiet.

Dade broke the silence. "Gov-ment men get done down there at the church house?"

"Oh. Yes, Daddy. They left a mess. Granny says yer to bring the mule an' yer ground-blade down tomorrow an' straighten it up. She says we need to plant new grass seed right away."

"You can help me, Geneva, directly after breakfast."

"I made my decision," Mattie announced.

"I thought you had," Dade replied.

What decision? Geneva wondered.

"We'll not tell the gov-ment about the baby's place

up there. He can lie here with his granddaddy. The final tide'll shelter them both."

Dade nodded agreement, then, seeming to realize what had been said, looked up from his plate. "What you mean, granddaddy?"

"You tell him," Mattie nodded to Geneva.

"Granny an' me knocked over Granddaddy's tombstone so's the men wouldn't see his grave. We rolled it over the hill into the woods. Granny said she'd promised to join him here, an' she didn't want him moved."

Slowly, a light moved into Dade's eyes. A brief chuckle soon turned into hoots and guffaws. Dade laughed. Great, loud roars of laughter caught Mattie into smiles and giggles. Geneva heard, and smiled her approval.

Dade shook his head. He swiped at the laughter tears in the corners of his eyes. "That woman. That daft, crazy old woman," he muttered. "Don't she beat all? She keeps Bart an' me at bay like some fox outrunning the hounds an' now she's outrun the whole United States gov-ment and outdone 'em all."

Geneva understood the depth of her daddy's laughter. It was his way of crying.

11

THE CHILDREN FLOPPED DOWN ON THE front porch, hot and restless. "What'll we do now?" Kendall asked. "We could play marbles, but Geneva won't. Come on, Gen. One game?"

"Naw. I ain't got no more interest in marbles. Anyway, it's too hot."

"Too hot to do most anything but sit," Alice sighed. "It'll be too hot to go to school, but at least it'll be something to do."

"Fourth of July tomorrow an' we ain't doing nothing then, either," James complained.

"Yeah. Remember last year?" Kendall sat up straight. "Remember? We went to town an' saw that parade an' heard them bands an' had ice cream an' everything."

"How come we ain't going this year?" Betty Lou said.

"Daddy said no," Alice explained. "He said there's so much to be done with crops an' moving an' all we ain't got time."

"I think my mamma an' daddy're too sad," Geneva

observed. "They don't feel like celebrating."

Kendall sighed. "An' school starts next day. I heard town kids don't go to school fer another three-four weeks. How come we got to start in *July*!?"

"Yeah. 'Specially when we're going to town school come September. Ain't no use going to that dumb old school *now*," James said.

"I want to," Betty Lou countered.

"Me, too," Geneva and Alice agreed.

"Town kids may not start school till August," Alice said, "but I heard they don't get out until *May*."

"I 'spect that's because they don't have no work to do," Kendall said. "Won't that be fine? To go to town school an' not have all this work?"

"I don't think so," James said. "I like farm work. I'll miss the cows an' mules an' even them dumb pigs."

"One good thing about school starting next day," Kendall said, "is playing town ball." He jumped to his feet. "Hey! Let's go down to Granny's an' see if she made us new balls fer this year."

Geneva looked at the shadows. It would soon be time to go home for chores, but she was supposed to check on Granny every day, and she hadn't been down to see her yet. She jumped up, too. "Race you," she yelled, getting a head start.

Granny sat on her porch, needle in hand and a pile of mending in a basket beside her. "I heard y'all coming way up the road. You ain't never gonna make you Indians the way you tear through the woods.

Come on up. Git you a chair. Geneva, get them young'ns cold water from the spring an bring me a new bucketful. Pour what's left to them chickens. Alice, look around the kitchen. Must be something y'all can eat."

"I found fried pies. . . . Oh, yummy, still-warm fried pies," Alice called.

"I guess I knowed you was coming." Granny smiled. "A body can't eat all them I made anyway. I just opened me up a blue jar fer a treasure an' out pops fried pies."

The children munched happily on the apple treat, still warm from Granny's skillet. "Umph, umph, Granny," Kendall mumbled. "You sure do make these the best ever. I think you better come to town with us an' make us fried pies every day. Bring yer blue jars."

"Now, don't you be a-teasing me, young feller." Granny frowned. "You think I'm talking daft an' I knows it, but I ain't. Just you wait till I die. Then you'll know about my treasures in blue jars."

"What kind a treasures you got?" Kendall asked, innocentlike, Geneva thought. He's trying to lead her on, but she'll squash him like a June bug.

"Don't you wish you knew," Granny said.

"Say, Granny," James interrupted. "We was wondering if you made us new balls fer the first day a school. We was talking about playing town ball."

"Oh, this daft old lady is good enough fer new balls"—Granny fixed her eyes on Kendall—"even if some people think she ain't got no treasures. Well, let me tell you, I got treasures in heaven an' treasures on earth an' treasures in blue jars an' only me an' the Lord knows what most of them is."

Kendall dropped his eyes.

"A new ball's a treasure. An' I used some more a my treasures in blue jars and I done made each an' every one of you children a new ball. New balls is treasures on earth." Granny reached into her mending basket. She rolled five balls across the porch where they bounced down the steps.

"Whoopee!" James and Betty Lou scrambled after the balls and tossed them to the other children.

"Hey, Granny, this is a good-un," Kendall crowed. "I believe this is the roundest and truest ball you ever made. Whatcha got inside here, anyhow?"

Geneva studied the heavy, round sphere resting in her hand. Granny's stitches, small and close, were nearly hidden in the circular folds of the intricate pattern. Geneva could count five separate pieces of old overall cloth that went together to make the sphere. Lord knows how many layers are inside this cover, she thought.

"Now I went down into one a my blue jars an' I found me five fine round buckeye nuts an' I wound each an' every one with all the string I been saving fer years. Yes sir, in my heavenly home I won't be needing that string. So don't none a you be coming down here to tie up no packages 'cause there ain't an inch a string in this house."

"Let's play catch fer a minute," James said. "Come on, Kendall, toss me one."

"First I want Granny to mark mine with my name. Those other kids at school will surely claim it once they see how good it is." Kendall turned to Granny. "Is there some way you can mark it fer me?"

"I shoulda thought that an' worked yer names into the cover 'fore I got it so tight." Granny cocked her head, birdlike, and peered at Kendall's ball. "I got one a them in-del-ible pencils. . . . Geneva, you look in that little drawer under my mantle clock. There's a bit a that inky-purple pencil in there, that kind that marks an' won't rub out."

Geneva opened the little drawer. She had found the pencil in the drawer once before but she hadn't looked at the rest of its contents. This time she looked.

First her fingers found an old key. Then a gold locket with a baby picture on one side and a wisp of baby hair on the other. Geneva thought it might be a Haw baby who'd died between her daddy and her Uncle Bart.

She also found a cameo brooch she had never seen Granny wear. She wondered if Granny had forgotten where it was.

"Geneva? Come on, Gen. Me an' James wants to play a little catch," Kendall called.

Geneva picked up the stub end of the indelible pencil and noted its point, carefully fashioned by hand with a knife.

"Geneva!" Kendall called.

Last of all she found a small envelope, yellowed and crisp with spidery, purplish writing. Could the letters have been formed with this very pencil? she asked herself. The letter was addressed to "Mrs. Augusta Kendall Haw." Surely not *this* pencil, but maybe—

"Geneva! Come on!" Kendall demanded.

"Ain't it in there, child?" Granny added.

"I found it." Geneva hastily closed the little drawer. I'll have to ask Granny about these things, she decided.

She presented the pencil to Granny, who held it carefully and pressed it against her lower lip. A purple smudge immediately appeared. Granny's wobbly hand worked out *Kendall Haw* and *1948* across the faded blue of the ball.

One by one the children silently handed Granny their balls. Geneva, as oldest, waited until last. "I'm not going to play with mine," she said. "It's all that's left of the old days, an' none of us will ever know how to make a ball."

James and Kendall tossed their balls into the air to catch them a few steps further on. Betty Lou, like the older girls, held hers quietly. "I don't know whether to play with mine or not," she said. "What would you do if you was me, Alice?"

"I think you ought to play with it, honey."

"Geneva, what do you think?"

"It's yers. You do whatever you want."

"Maybe I can do both, play with it some an' keep it too. Oh, Granny forgot to make Sissy one. An' Tad."

"We'll have to ask fer two more," Geneva agreed. "Even little brothers an sisters need a granny-ball."

"I better hide mine," Betty Lou observed. "Tad'll want it an' Mamma'll think I ought to give it over 'cause he's younger."

"I'll see Granny tomorrow. I'll ask her to make one fer him right away, so you'll only have to hide it a day er two. Anyway, Tad can't have it, it has yer name on it."

That problem settled, Geneva and Alice turned their talk to the new year of school. Kendall interrupted: "Gen, you an' Alice an' that Paul Gordon are the biggest in the school. Which of you two's gonna be captain fer the town-ball game first day?"

"I ain't." Alice answered. "I don't like it when that ball heads fer my feet er legs. It *hurts*."

"That leaves you an' Paul Gordon. How about it, Gen, you gonna be captain? If you ain't, it'll be me."

"What about them other boys in yer class?"

"They don't amount to nothing. I'll captain one team 'less you take it."

"Reckon I will. I been waiting eight years to captain the town-ball team." Geneva's voice held a tease Kendall failed to hear.

"Aw, Gen, what'd a girl want to be captain for? Boys er natural captains; girls ain't."

Geneva bristled. "Listen, smarty, captains of teams

is neither. Teams get captains fair an' square by who's the biggest an' who's in the biggest grade. If Alice don't want it, we won't be drawing straws fer the other one. You have to wait, same as I did last year, though I'd made a better captain than that Obed Drews."

"Come on, Gen," Kendall begged. "This is our last year at school. I'll never make captain once I go to school in town."

"Kendall, shut up. It's Geneva's turn fair an' square," Alice interrupted.

"How do I look, Mamma?" Geneva asked the first morning. She didn't have a new dress for the first day, but last year's Sunday dress, with its hem let out, would serve.

"Well, you'll do." Mattie cast a last look at the cut hair. "Can't you do nothing about that gouge in that fringe?"

Geneva turned to the mirror over the washstand. She parted her hair further down and tried to pull it over the short-cut hair. "That's where I sneezed." Geneva grinned.

Mattie went into her room while Geneva fiddled with the sticking-up hair. When Mattie returned, she handed Geneva a curved comb with a fancy edge. "Here. Here's a comb I had when I was a girl. Try putting it in there somewhere."

Geneva looked at the comb, topped by a swirl of design. It was neither brown nor black but shone with an amber glow. "That's pretty, Mamma. What's it made of to look so shinylike?"

"That's celluloid. Now, what you do is smooth the hair back an' put the teeth into the hair from behind. See? You practice there a minute an' you'll get the hang of it. Why, when I was a young'n we done our hair in what they called rats—big, puffy things that rolled yer hair way out. They was heavy on yer head an' you got so tired carrying it around. . . . I expect that's why Granny cut her hair off, an' probably you," Mattie added. Geneva nodded. The haircut fight was over.

Geneva picked up her basket of lunch, gathered up the books she had borrowed at the end of the last term, waved good-bye from the back door, and took the high road to her cousins'.

Alice and Betty Lou waited on the porch. Tad danced around them. "Geneva, they won't take me. You take me. I wanna go to school, too. James an' Kendall went without me. You take me."

"I wish I could, Tad," Geneva hugged the child, "but you just ain't big enough yet."

"Mamma, we're going," Alice called into the house.

The first day of their last year began like any other first day: with excitement, dread, and eager reluctance.

As the biggest in size and the oldest in grade, Geneva, Alice, and Paul Gordon knew to sit in the big desks at the back of the small room. Arranged in front of them were the boys in the seventh grade, and so on, until, in the very front, were the little ones just starting first grade. Miss Anna Belle Porter was their teacher, as she had been for a number of years, so nearly all the children knew what to expect. She made the same little welcoming speech as always and went on with the announcements.

"The school trustees have decided to hold school until all of you have left the river bottom. They would like to hold school until December, though they know many of the families will be gone by then. That will give some of you nearly a six-month term for this year. The school, along with all the rest of the buildings—farms, homes, church houses, all—must be razed to eighteen inches by January, 1949. The trustees decided it would not take long to raze this little building. We will have school as long as we can. Some of you, who are not moving beyond the Ridge, can perhaps come back to finish the term.

"I know this sounds pretty upsetting and worrisome, but we will try to carry on each day. We will not let a thing like a tide of water keep us from learning all we can. Any questions?"

Betty Lou stood up. "Yes, ma'am. What do you mean, raze? I thought that meant to, er, puff something up, like 'raise them biscuits with self-raising flour.' "

The children and teacher laughed. Miss Anna

Belle explained the difference between "raise" and "raze."

"Now, about our school year," the teacher continued. "Geneva, your main strength is arithmetic. You will be the teacher of arithmetic for all the children in grades one and two. I'll help you plan the lessons, of course. We can do it each day at lunch time. Alice, your main strength is spelling. You'll have charge of the spelling for grades one and two. Paul Gordon, you will assist me with geography. The boys of the school will see to the fire when it is time for the stove, and all of you are expected to teach the younger children the games we love to play."

As the teacher's voice went on and on, Geneva allowed her thoughts to consider her work for the year. Arithmetic pupils! All hers to teach. Why, she was near as good as a teacher already. Granny would be so proud, an' so would her mamma. Oh, she had often helped with the lessons; all the pupils did, but for her to be the real teacher for arithmetic was special.

Miss Anna Belle's voice penetrated Geneva's thoughts. "And now, as is our custom on the first day of school, it's time to choose captains for the town-ball game. Paul Gordon, after the teams are chosen, you select a slab of wood from the woodpile for the bat."

"I already done that, ma'am. I brung two from home. I brought a good heavy one fer us big kids an' a smaller one fer them others. See, I whittled them a handle."

"You *did* that, Paul Gordon, and you *brought* the wood from home," Miss Anna Belle corrected.

"Yes, ma'am, I already done it. See, I made them little kids a handle on theirs."

Miss Anna Belle sighed. "That's fine, Paul Gordon. Thank you. Now then, between Alice and Geneva, who's to be our other captain? Have you girls settled this? Who is the older?"

"I am, ma'am," Geneva said.

"I don't want to be captain," Alice added.

"Geneva, then?"

Geneva caught Kendall's imploring eyes. His mouth formed a silent "please." Geneva slowly rose to stand beside her desk. "I . . . I choose Kendall Haw to be captain." She sat down. Alice squeezed her hand. "I didn't do it for Kendall," Geneva whispered. "I did it because—well, because if I'm the arithmetic teacher for five little children . . . I don't think teachers play town ball. I think they only let the kids play. Besides, I want to keep my granny-ball forever."

Anyway, she thought, things er too sad right now fer playing.

12

THE ONLY PART OF THE DAY GENEVA could keep her mind on what she had to do was when she worked with the younger children and their arithmetic. It wasn't the beautiful fall weather that occupied her thoughts but the changes happening so fast—too fast—to her world.

In late August crews of men with saws, big log trucks, gasoline to set fires, and steam-driven earth-moving machines swarmed over the mountains that circled their river bottom. They cut every tree that might be in the future lake bed. They hauled off the big logs, piled up the brush, and burned piles of it night and day. The air was smoky and hazy from the fires, and the barren hillsides looked naked and ugly.

In September the tearful families separated as Bart, Ina, and the children loaded the last of their goods into their wagon and truck. Now Geneva walked to school alone.

Dade and Mattie worked from first light to long

past dark. Mattie organized and Dade hauled the household and farm through the river bottom, up the long climb to the Ridge and their new farm. Dade complained that he was either coming or going, never staying still. In between the massive, wrenching work of moving, he cut the last hay crop and corn, made the silage, and hauled that to the new place.

Mattie and Geneva closed up their garden, canned what needed to be saved, dug potatoes and carrots, and sent the cabbage up to the new cellar. Mattie fussed. It seemed to Geneva she could do nothing right, especially the packing of the already-filled jars of food. "Mind those jars er wrapped good," Mattie said again and again. "Wouldn't this be a mess should they break? What would we eat if that should happen?"

"Yes, ma'am," the girl said, stuffing more straw around the jars in their packing crates.

"You better stay home an' help yer mother," Dade said one morning. "No use you going no more."

"I have to go to school, Daddy. Miss Anna Belle is counting on me. I'm the arithmetic teacher."

Dade grunted. "She's going," Mattie declared.

I'm going, Geneva thought. Come what may, I'm going to finish grade eight. I might not get to high school, but I'll finish this.

Little by little the farm and house were stripped and hauled off. The house was almost bare, but they were reluctant to make the final break.

In November men came and took down Bart and Ina's house. It had been sold and the new owners prepared it to be moved.

"Is someone going to take our house, too?" Geneva asked.

Dade shook his head. "No. I'll take it down. Part of this house was built when Daniel Boone first come through here, an' if it can't be a Haw's house, it ain't gonna be nobody's house."

Geneva watched the destruction of the woods and forests she loved so dearly. Men and saws went higher and higher up the rise of mountains. They went beyond the baby's burying ground, past the woodlot, above the high pasture Geneva had taken the cows to before the tide of last spring. Castle Rock was left exposed to the wind and rain. Across the river the setting sun no longer danced on Tillet's Mountain but burned red through the haze.

Rain came and created gullies in the thin soil. Jagged rock fell off exposed cliff faces. The gold of the long fall weather turned to the dusty, smoky gray of fires, spoiled land, destroyed lives.

Had it not been for her pupils, Geneva would have listened to Dade and quit school then and there.

The government men made repeated trips to Granny's. They took down the church house and took the trees from the woodlot behind the burying ground, but they left Granny, legs spread and shotgun in hand, standing on her porch.

"What'er you gonna do about that woman?" they pressed Dade.

"I reckon the problem'll be solved one way or the other," he would answer, turning away from their demands.

At school the pupils dwindled. It seemed each week saw them saying good-bye to yet another family. By November, Geneva had only two little arithmetic pupils and she had taken over their spelling, too. Even Miss Anna Belle, usually so spirited, seemed dejected and quiet.

"Let's have our eighth-grade graduation," teacher said one morning. "You're here, Geneva, and so is Paul Gordon. I've asked your daddy, Paul Gordon. He's the only trustee left, and he agrees. We'll have it December first and that will be the last day of school."

Mattie ordered a new dress for Geneva, and new shoes. Geneva learned two recitations and the few remaining children halfheartedly learned some new songs. They used up the cherished art paper to decorate the room, which they cleaned to a shine, and on a Friday afternoon a handful of parents arrived for a sad and tearful graduation ceremony. Miss Anna Belle gave the children all the books, the crayons, the pencils, the maps, the worn and well-used materials. The little building was left empty.

Three days later, the men of the community took down the walls, carried away the windows, gave away the desks, left nothing but the stones of the foundation and the two steps.

"We'll have to move next week," Dade announced at supper.

Mattie nodded in agreement. "What about Granny?" Geneva asked.

"I been studying on that," Dade answered. "I reckon we gotta leave her here as long as we can. None of us has been able to get her to see reality. Maybe the tide rising around her doorstep'll do it."

"But . . . but we can't just go off an' leave her all alone—alone in this whole river bottom! We're nearly the last family to leave." Geneva was outraged.

"I thought maybe I could get somebody to stay with her, like Sheffield's girl, that one that ain't just right in the head," Dade said. "If Granny'll have her."

"No! No, I'll stay with her. She'll listen to me. You an' Mamma go on to the new farm. You'll have plenty to do up there, but I'll take care a Granny as long as—she says she's gonna die before the tide comes. Let's give her time."

Dade studied Geneva's face. "I think yer right."

"I'll take books, an' quilting, an' crocheting, an' I'll see to her being warm, an' maybe I can get her to move."

"I can come by every day er two to bring you what you need," Dade said.

"What about the milking? Mamma can't do it an' we got all a Bart's cows."

"Reckon I can find a feller up on the Ridge to come work fer me. Woodcutting's all done. There'll be plenty a men looking fer work this winter."

"Let the child take a mule, an' that little jolt wagon," Mattie said. "Then if she needs us, she can come get us."

Geneva loaded things into the jolt wagon and headed for Granny's. The old woman met her at the door, shotgun resting over her arm. "I heard you coming. I know'd it wasn't them gov-ment men. They come in trucks, but they could be fooling me with a wagon an' a mule. What you doing here, child?"

"I come to stay with you."

"Why? I don't need you. I can take care a myself an' them fellers, too."

"I want to."

"Oh. Well, unload that plunder an' come on in. It's cold out here today an' I don't want to look at this plundered place.

"Open up that little room." Granny pointed to a closed door. "Ain't nobody slept in there fer years. Reckon it'll be dusty, but you can light in there."

Geneva and Granny settled in.

On Christmas Day, Bart, Ina, and the children arrived in a fine new car. "Business is good," Bart crowed. "Stoves an' refrigerators an' vacuum cleaners an' all that stuff sells like hotcakes."

The cousins wandered among their destroyed world. "This is where . . ." "Remember how . . ." "Our barn was . . ." "What happened to . . .?"

They wandered the empty spaces. They found the doodlebug rock, now looking like any other fallen piece of cliff. They climbed the washed-out gullies to Castle Rock, then hid their eyes from seeing the barren mountains and once-proud farms below.

Back at Granny's, they sat among the grown-ups, who talked in low, sad voices; "I heard up town the dam's all done an' the gates are shut," Bart said. "Two weeks early. They say it'll take a year er two to fill up the lake, but reports are it's filling already."

"Been a lot a rain this fall, an' Lock Twenty-one is full. They're to blow her next week, that'll fill the lake up plenty," Dade speculated. "Tide'll rise fast. Mamma, what you aim to do about moving?"

"I ain't."

"Them gov-ment men'll move you if we don't," Bart said.

"I'll take the shotgun to you same as to them." Granny's voice was matter-of-fact.

Bart laughed. "I reckon you would."

Granny nodded.

"Seriously, though," Bart said. "You gotta make a plan. Now, you can come up town with us, we'll give you a room all to yerself, er you can go up to Toller's Ridge with Dade. Which is it to be?"

"Neither."

"Mamma!" Bart exclaimed. Dade said nothing.

Geneva knew he had had this discussion with Granny many, many times. Geneva doubted Bart would get any farther than Dade had. "Mamma, what you gonna do when the tide rises around the porch, there, an' slips under the door?"

"I'll be dead by then. Geneva an' me's gonna dig me a hole over there in the burying ground and that's where I'm to be layed. I'll have it all ready. You don't need to call in any preacher or nothing: don't make a fuss about it so them gov-ment men don't find out nothing. Just lay me there—don't need no box er nothing. You do like I say."

Geneva saw Dade and Bart exchange a long look. Bart's seemed to say "like hell we will," while Dade's seemed to show he was more at ease with Granny's plan.

The day after Christmas, Granny directed Geneva in the digging of a hole for her burying. Geneva protested, but it did no good. The hole was dug.

Winter settled in. A winter of unprecedented rain, a winter of several deep snows. The snow hid the destruction of the mountains. It prevented Geneva and Granny from seeing the future looming ever closer. Geneva brought in split wood, carried water from the spring, helped Granny butcher the last of the chickens, fed the mule, and tried not to look beyond their immediate view. Granny let Geneva manage their lives and refused to leave the house unless it was necessary. They spoke little. Geneva supposed Granny was grieving, too.

Dade traveled between Toller's Ridge and Granny's cabin, stretching the days between visits from two, to three, to four as he was assured by Geneva that all was well.

Reports of the water's rising reached Geneva and Granny as January turned into February. From the base of Toller's Ridge they were only a scant five miles from the dam.

Early March saw the first greening of the pastures and fields, and weeds quickly filled in where once proud corn and tobacco had grown. Geneva found wildflowers on the low slopes, but spring's return brought Geneva and Granny little pleasure.

The government men appeared every few weeks. Geneva met them on the porch. She learned to say what they wanted to hear. "We're getting ready to move." "Soon." "A week er two." Granny and her shotgun waited behind the closed door. The men knew it, and respected Geneva's word.

Inside the house, Geneva wrote down all of Granny's stories. She read Granny all their books, read again and again the Bible passages the woman loved, helped piece together one more quilt, and allowed Granny to tease her about her funny, crooked stitches. She wrote letters to all of Granny's relatives who had long ago emigrated to other parts of the country. She pinned names on quilts and keepsakes.

Geneva, Mattie, Alice, Sissy—all would receive something. Dade, Bart, and the boys were to be given things left from Granddaddy.

One Wednesday afternoon Dade came with his truck. Granny met him at the door. "What you doing here?"

"I come to see you. I brung you some meal an' flour an' coffee. Mattie wants yer washing."

"You got something else in mind. Won't do you no good. I ain't going."

"But Mamma, I can't let you stay here an' drown. What'll folks say about that, that yer own son Dade Haw left you to drown?"

"They'll be saying what a good son you was to let yer mamma stay here an' die an' be buried by her man in her own churchyard. They'll be saying that 'cause that's the way it's a-gonna be."

"Don't seem to me yer gonna die any time soon." Dade's voice was weary. "Yer making me an' Mattie an' Bart an' Ina a big problem, Mamma. You an' your stubborns. The sheriff's been to see me, and he's gonna come get you himself if'n you don't go off with me."

"I can shoot that sheriff, er you, er any of 'em. An' I will."

"An' you'll get dragged off to jail like a common criminal. They'll hang you an' you won't come near this burying ground, an' you'll go to hell besides."

"Oh."

Geneva listened to her father, eyes round with the words he said. Her soft-spoken daddy was seldom so direct, especially with his mamma. Would Granny be persuaded by his words and his tone?

Granny stood up. She pulled herself straight. She looked Dade in the eye. "I'll take my chances with the Lord. He's in charge here, not you. I gotta wait on him. He'll let me keep my promise to my man, don't you worry none. I'll be dead an' buried afore that tide gets here, you just see if'n I don't."

Granny turned, marched into the kitchen and returned in a moment. She had Granddaddy's gun. Geneva could only hope the old woman hadn't found the shells she had so carefully hidden. "I aim to be here when my Lord comes after me. You nor nobody else is going to take me away, you hear me, boy?"

Geneva looked from her grandmother to her father. Dade was the one to drop his eyes, so fierce were Granny's black ones and so firm the set of her toothless mouth.

"Now you get on outta here an' tell them others I ain't going. The Lord's the one who's gonna move me, not you er them er that sheriff nor nobody. Here me? Git!"

Dade rose, giving in to weariness. Geneva thought, my daddy's old an' small. Dade sighed and went out the front door. Geneva followed, longing to put her arms around him, knowing she could not; he would not allow it.

He pulled himself into the truck. She saw tears in his eyes and dropped her gaze. She did not want to see him cry.

Dade looked across the ruined river bottom. "Geneva, if she ain't dead or moved by Sunday, I really will be back with the sheriff. Maybe he can scare that stubborn old woman an' we can get her off. You gonna be all right with her until Sunday, child?"

"I'll be fine, Daddy. She don't scare me with her gun. Anyway, I got the shells hidden. Don't worry about me."

"Of course I'm gonna worry, child. Now, you got the mule. If you need to get us quicklike, get on his back. He'll run if you poke him a good one. We'll be here Sunday after church, all of us. She'll go, one way or the other."

Geneva watched his truck until it was a mere moving shape far along the road to Toller's Ridge. It seemed the only moving shape in the whole of the river bottom. She reentered the cabin, knowing full well what her granny would ask.

"He's gone? What'd he tell you, girl?"

"He said all of them er gonna come back here Sunday after church an' you better be ready to go er else."

"Else what? Ha! He ain't in charge of the else what, is he now?"

Geneva had no answer. She could only wish that somebody was in charge of the "else what" beside her and the Lord.

13

THE RAIN THAT STARTED SHORTLY AFTER Dade and his truck disappeared up the mountain kept Granny and Geneva in the house all of Thursday. They were mostly quiet, Granny calm and Geneva edgy. She watched the clock and wished the days until Sunday would hurry along.

Friday morning the sun rose high into a blue sky, and the few meadowlarks still nesting in the pasture fields welcomed the pretty day. The sun drew Granny and Geneva to the back porch. Both noticed a sparkle below that had not been in sight a few days before.

"Don't that river seem uncommon wide? Has Bart forgot to open Lock Twenty-one this spring?"

"Lock Twenty-one is gone, Granny. Remember? The gov-ment blew it up two–three weeks ago. That's the tide. The final tide. It's come to claim this land, just like we been telling you."

"Oh."

Granny closed the door. She sat heavily in her rocking chair. "Then the time has come."

"Granny, what you gonna do?"

"Surely the Lord will come fer me now. I'll make myself ready. He'll come." Granny bowed her head. Geneva stood silent.

"Geneva, you listen now, an' do like I say and write down what I tell you so it will all go the way the Lord an' I wants it to go.

"Build me a nice fire, an' bring me some water. I'm gonna take me a bath an' wash this blessed hair that's a-growing already."

The bath prepared, Granny turned to Geneva. "Well? You gonna stand there an' watch an' old woman naked?"

"No, of course not." Geneva retreated to the front room.

"Leave that door open so's you can hear me. He might come before I'm ready. Don't you listen to me 'less I call you, child. What I'm a-doing ain't yer concern."

Geneva, reeling from the conflicting orders, pulled a rocking chair to the far side of the room.

"Get you the Bible, child. Read me some a them Psalms. The ones that tell about the flood."

Geneva opened the book and began the reading, but her mind could not take in the words Granny wanted to hear. From the kitchen Geneva heard Granny sigh as she eased herself into a chair. Geneva paused in her reading. "I'm a-coming, darlin'," the old woman called out, and Geneva realized she was talking to the memory of her husband. A shoe dropped to the floor. Granny's quavery voice rose in an old hymn tune.

"Come home, come ho-o-me. Ye who are weary come home. . . . Lord, I'm weary. The flood's rising, Lord, just like you said it would. I'm a-fixing myself to come home."

The other shoe dropped. From the kitchen, a long silence, then, "Bath water's nice an' warm. Oh, Lordy, it'll be good to rest in His arms. I gotta hurry, Lord, the water's rising around my door an' my man's there in our heavenly home, waiting on me. I'm coming, darlin'.

"This here's the last piece a my lye soap, Lord, so don't leave me no longer." Granny sung the next few lines of the hymn as she swished a cloth through the water: "Earnestly, tenderly, Jesus is calling, calling oh, sinner, come home. . . .

"I'm gonna lay me down on that there table, Lord. You'll find me there, waiting on You. I'll be all ready fer my burying. Them gal's only have to sit with me tonight an' then them boys can put me right in that hole me an' the child's readied.

"Lord, You are my shepherd, no want shall I know but to come home. I won't lay me down in yer green pasture, Lord, but here in my little kitchen an' you can come in yer own good time, but don't be too long about it, Lord. That water's rising mighty fast."

"Geneva, I don't hear you reading."

"I am, Granny, honest. I was just ready to start the Ninety-third Psalm. . . . Oh, Lord, to whom vengeance belongeth . . ."

"Geneva, bring me them clothes laying on the top a my quilt chest."

Geneva rose to do the old woman's bidding. She better understood, now, what Granny had in mind. Geneva peered into Granny's eyes, wondering again if the old woman was daft or if she should fetch Dade and Mattie.

"What you looking at, child? Stop yer studying me. I ain't daft. I'm getting ready fer my Lord to take me home.

"Now, I want that blue dress with the white collar folded inside that chest.

"You empty out that bathwater, child, and ready up this here kitchen. It ain't hardly fit fer the Lord to enter."

Geneva followed Granny with the corner of her eyes. She pulled the galvanized tub through the back door. Granny seemed her usual self. She pulled the dress over her head, buttoned the front with firm fingers, hummed as she sorted through her pile of clean aprons. She sure doesn't look like she's about to die, Geneva thought.

"Don't reckon I need me an apern," she muttered. "Surely my Lord'll know me without my apern." Granny giggled. "But my man, oh, my darlin' man, *he* won't know me without an apern." Granny chose an apron, pulled it around her waist, shook her head and folded it back up. "Ain't never seen no one layed out with an apern on. Reckon that's kind a silly."

Geneva stood outside the kitchen window until Granny caught sight of her. "I got something more fer you to do, child. Come on in here now! What you standing out there fer anyway? Ain't you never seed an old woman get herself ready to meet her Lord?"

Geneva entered the kitchen. She couldn't contain a smile. "I don't believe I ever have, Granny, have you?"

"Now, Geneva, after the Lord takes me, you tell Mattie an' Ina I already done bathed. They won't need to be laying me out er nothing. The hole's ready. You get that quilt you an' me done finished last week. That's to roll me in fer the hole.

"Open up that table, there. I'll lay myself out on that table. Then nobody'll have to move me.

"You know what I want from that Bible, child. You tell Alice an' them little fellers to read one each. They ought to know them by by heart, but I don't suppose they do."

Geneva interrupted. "What about this table, Granny? What do you want me to do?"

"Open it up. Spread them legs apart an' put in them other leaves. I's longer than that table. Them boards er under yer bed."

Geneva brought the boards and put them in the table. She wrote her name in long-settled dust and added hearts and arrows. Granny giggled. "Pshaw, child, I ain't seen me one a them hearts since I was a girl in school, 'bout yer age. I seen me some then."

"Who sent you hearts?" Geneva asked, wanting to distract the old woman.

Granny's face flushed. "Why, I remember this boy and that boy.... but the only one really mattered was yer granddaddy. Yer granddaddy's waiting fer me an' I'll soon be there to see him."

"Did Granddaddy ever write you a love letter?"

"One. Only one. I kept it forever, but I ain't seen it fer years an' years. Oh, that man had a way with words. He could a been a real poem maker if he hadn't worked all his life. Why, he'd recite me poems he made up, but I never had time to write them down.

"I'm coming, honey. I'm coming soon." Granny crooned to the sky.

"Wouldn't you like to read that letter again, Granny? Can't you remember what he said?"

"He had such pretty handwriting. We was taught good handwriting in my school. Not like now. Oh, he could make them curlicues, an' them floating ends. Why, he wrote out my whole name, Augusta Kendall Haw. Ain't that pretty? Augusta Kendall Haw. Ain't nobody calls me Augusta an' it a pretty name."

"Granny! Granny, I bet I know where that letter is!"

"Sakes, child, how'd you know that?"

"Come on. I'll show you!"

Granny followed Geneva into the big room. Geneva pulled open the little drawer under the clock.

"Sit down. Here." She lay the drawer in Granny's lap.

"Why, I swear . . . don't that beat all? I been wondering where them things was. I been studying it fer years an' years. Why, here's the baby girl, my only little baby girl. Here's her picture an' a bit a her hair. Oh, Lordy. I'll soon see my baby girl in my heavenly home, too." Granny rocked the locket against her heart. "Put it around me, gal. Then my baby girl'll know it's me. Oh, I was young then. I'm old, now. She'll need help knowing her mamma in our heavenly home."

Geneva worked the old clasp and tucked the chain under Granny's still-damp hair. "An' here's my granny's cameo brooch! Ain't it a pretty thing? I ain't worn it forever." Her old fingers opened the pin. "Reckon it'll look good on this here collar. No. I can't take this. This is a treasure on earth."

She looked up. "Now, child, you remember them treasures in blue jars. Don't you be leaving after my burying without them blue jars."

"What about the letter, Granny?"

"I see it. Plain as anything. I see it. But I ain't done with this cameo brooch. You have it, child. She what owned it was yer granny from my granny. She was the one part Cher-o-kee Indian. She give me my dark eyes, an' you yer dark hair, so yer to have the cameo brooch."

"Thank you. I'll keep it forever an' tell my girl—

when I have one someday—where it come from."

Geneva accepted the pin. "What's that key, Granny?"

"Why, it's nothing but an old key, child." Granny's face flushed again.

"Why'd you keep an old key if it's just an old key?" Geneva teased.

"He gave it to me. Yer granddaddy gave it to me. He said it was the key to his heart an' I was to keep it forever, for I'd have his heart that long."

"Why, Granny, that's right nice."

"I had it forever, but I didn't know where it was. I reckon I'll put it in a handkerchief an' take it with me to my burying."

"What about the letter?"

"Yes. Here it is." She held the fragile, yellowed envelope in her hand. "I can barely read it. I can hardly see it, but look, there's my name: *Mrs. Augusta Kendall Haw.* Ain't that a pretty name? Oh, he'd call me Gussie same as everyone did, but when he spoke nice to me, he'd say 'Augusta' so pretty, an' his voice would tell me how he was a-feeling."

"Are you going to read it again?"

"Don't reckon I can. You read it to me."

Geneva took the fragile paper. The envelope crumbled as she carefully withdrew the single folded sheet. The fold tore along the edge, though she opened it carefully. Pale lavender, spidery writing slanted across the paper, but so pale were the curlicues Geneva could not make out the words.

"Here, let me go close to the window.

"It says 'My dearest Augusta . . .' "

"Yes. I know that."

"I . . . I'm sorry. I can't make out another word of it. I'm sorry, Granny." Tears filled Geneva's eyes. Tears at the loss of these words of love, words that would have meant so much to an old woman.

"I can remember them in my heart," Granny said. "I can't tell them to you, but they are in my heart."

14

GRANNY SIGHED A LONG SIGH. "THROW that thing in the fire," she said sharply.

"What?" Geneva said, startled out of her reverie. "Throw what in the fire?"

"That letter. Let it go."

"Oh, no, Granny, let me keep it. I want it. Don't burn it up."

"What you want it fer? Can't read it."

"But it's from Granddaddy. I can barely remember him an' I like to think about him writing it."

"Well, it don't matter to me, I reckon."

Geneva turned toward her room. She would put it in her own Bible, that's where people kept precious things. "Anyway, it's a treasure on earth," she said.

Granny rose from her chair and moved toward the kitchen. When Geneva found her, Granny lay on top of the table, stretched to her full length, eyes closed but a frown creasing her forehead.

"Granny! What are you doing?"

Granny sat up stiffly. "Table's durn hard. Get me

that quilt, child, an' a pillar. 'Pears like I got me a long wait."

"Granny, *nobody* gets on the *table*. Get off there, now." Geneva was growing tired of humoring the old woman. It seemed time to try something else.

Granny slid off the table. "Git me that quilt. Ugly old thing's good enough. Git!"

Geneva brought the quilt. Granny spread it over the table, first with backside up, then turned so its pieced pattern was on top. She brushed at the dust picked up from where Geneva had drawn the hearts. "Now, how'd this here new quilt get dirty?" she muttered. "Well, it don't matter none. It's about to get dirtier."

"What'er you saying?"

"Nothing. Git me that piller off my bed. That high, fluffy one I like."

Geneva fetched the pillow. "Now, help me up here, child. I won't be getting off again."

"No."

Granny's black eyes snapped. "If you ain't gonna help me you might as well hitch up that mule to that jolt wagon an' get on out of here."

Reluctantly, Geneva helped Granny back on the table. "Smooth down my skirt, child. Tie up that shoe." Granny arranged her hands on her chest, composed her face, sighed a long sigh. "I'm ready now, Lord," she muttered.

"What?"

"None a yer business. I'm speaking to my Lord."

"What you aim fer me to do?"

"Whatever you want. Wait on the porch, I reckon. Look at what used to be our pretty place."

Geneva stood in the doorway. She watched the old woman breathe deep, then deeper. Granny appeared to be asleep. Asleep? Geneva leaned close. The breathing was regular, Granny *was* sleeping. Likely she thinks she's dead, Geneva thought. She looked at Granny's relaxed face, mouth slightly open to a small, snuffly snore, wrinkles eased around the eyes, hair spread on the pillow. Granny's color was good and rosy, not "dead" like that of kinfolks Geneva had seen laid out.

Geneva studied the woman. I gotta think a something to bring her back to knowing she's alive. There must be some way.

In the silence of this whole blessed world, Granny's clock ticked away. Geneva sat in Granny's rocking chair to think. She may have dozed, too, but the next sensation she had was of hunger. It was near dinner time. I'll make something good fer dinner, something Granny really likes, like . . . greens. We ain't had salat greens all winter. Geneva quickly took a sharp knife and basket and tiptoed out the back door. Surely among all the destruction of their world she would find pokeweed or dock or early cress or even dandelion. Greens would prop Granny up. Granny had always told her salat greens were a spring tonic. Well, let me physic that old woman till she can't lay down fer trotting to the outhouse, Geneva decided.

Knowing where to look, it took Geneva only a few minutes to find a nice basketful of salat. She had also planned what she was going to do.

She strode back up the steps, across the porch, gave the door a good hard swing and slam, and banged open the stove. She caught Granny's eyes following every movement; but when she looked, Granny pretended sleep.

Geneva quickly built up the fire, put the iron skillet over a burner, and scooped in a generous spoonful of bacon grease. She rummaged in the potato basket, found nothing there, so again slammed out to the potato cave. While there she took a head of cabbage. She wanted this dinner to smell to high heaven and cabbage cooked with a good stink.

Back in the kitchen, the grease in the iron kettle was sputtering, so Geneva washed the greens and let the water sizzle into the heat. Maybe she could scare Granny into thinking the house would catch fire.

She cut the potatoes into another skillet, threw a split of wood into the stove, and banged on another kettle of water. She would boil that cabbage and swim it in butter. Still Granny pretended sleep.

Geneva brushed off the quilt on one side of Granny and laid plate, cup, and silverware. She stomped to the other side and fixed another place. She pushed at Granny's hip to find a place for the cup. Granny's eyes struggled to stay closed. Geneva turned back to the stove. The fire was getting so hot she feared potatoes and greens would burn. She opened the damper, sending some of the waves of heat up the chimney. She could feel Granny's eyes studying her back.

Geneva decided to fetch some new water. That'd give Granny time to look the dinner over, for the smells of the potatoes, the greens, and the boiling cabbage were making Geneva mighty hungry. Surely Granny, too, was feeling the power of being alive and hungry.

Bucket full, Geneva took a deep breath. She had made a decision. She was not going to let Granny die. It seemed to her the Lord wasn't ready to take Granny, anyway.

That's *it!* thought Geneva. That's how I'll operate on Granny. Time fer a little playacting. A lot of playacting.

Geneva strode into the house. "Well," she demanded. "You still alive?"

Granny's eyes flew open. "Fine way to talk to a woman waiting on her Lord."

" 'Pears to me you got a mighty long time to wait."

"Why? I'm ready."

"You may be, but the Lord sure ain't."

"What'd you mean?"

"I mean it 'pears to me *you* are making this decision, not the Lord. You er the one's decided to die, the one trying to tell Him what to do. You ain't no more ready to die than the man in the moon."

"I am too," she whined. "What do I have to live for? Seems to me He's plainly telling me that I am a-going to my heavenly home. That tide's coming to take my house an' my fields, an' my church house is gone. My sons have gone an' left me an' my man's laying over there waiting. That's the Lord's way a telling me I'm next."

"That's yer decision. It sure don't seem to be His."

"Hmphh." Granny lay back down.

Geneva moved the greens off the fire with a mighty clank. She poured out the morning's coffee, added new grounds, and set the pot to boil. Granny loved her coffee . . . and the smell of new coffee. Granny wiggled a little but kept her eyes squeezed shut.

"Granny, yer in my way. I got the oven hot an' I'm making corn bread. It's mighty hard to hold these pans in my hands an' stir in that grease, too."

"Hmmmph," Granny muttered.

Geneva slid the pan of corn bread into the oven. In a few minutes, that would be done. Was she rousing Granny? Geneva was uncertain.

The coffee began to boil. Geneva pushed at Granny's skirt to set down the butter, molasses, and applesauce. "I wish you'd get off there in that good dress. I'm afraid I'll get butter on it er something. Don't reckon the Lord'll want to see you in a spotty dress. I'd say He don't want to see you *no how*."

Geneva strode into the big room. She had a minute or two before the corn bread would be ready. Let the old woman think.

A quavery, teary voice let her know she had won the battle. "Geneva?"

Geneva strode in. Granny was sitting on the edge of the table. "If I'm making the Lord's decision fer Him, what do you think He aims fer me to do if it ain't to die?"

"I think He wants you to get off that table an' eat this dinner I made."

Geneva reached out a hand to help the old woman into a chair. She put a cup of coffee in front of her. Soon the table was filled with dishes and bowls of the good dinner. "Yer setting that stuff right on that quilt," Granny whined.

"An don't it make a pretty table cover? Better'n a winding sheet. All winter you planned it fer a winding sheet, didn't you?"

Granny nodded. "I thought I'd be dead by now."

"Well, just goes to show you what happens when a fool of an old woman tries to tell the Lord what to do. Who do you think you are, telling the Lord you are ready to die? What gives you the right to decide something He ain't decided? *Now eat!*" Geneva commanded.

Granny peeked into a bowl. "Them's greens, ain't they?"

"Spring tonic. Get 'em on yer plate an' into yer

mouth. You know there's strength in spring greens. Strength fer the days ahead."

Geneva watched Granny's plate. When the old woman wavered, Geneva plied more food. "I'm full to busting," Granny complained.

"Good. You ought to be. Now, Granny, I made the dinner. It's only fair for you to do the dishes. I got things to do in the other room." Geneva was amazed to hear herself give orders, and more amazed to see Granny carry them out. "Here's yer apron."

Geneva took the broom to the big room. She made a fuss with chairs and rugs.

"What you doing, child?" Granny called.

"Spring housecleaning. This place ain't been cleaned forever."

She threw rugs onto the front porch. "An' when yer done with them dishes, I'll be ready fer some help here. We're going to wash these dirty windows inside an' out."

By midafternoon the little house was clean. "Come on, Granny, let's sit on the porch, like the old days."

"No. Can't stand that front porch an' what it looks like out there. Let's sit on the back porch. We can look at them fields. Anyway, there ain't nobody coming along that road."

Geneva brought a piece of cold corn bread and a cup of cold coffee for each of them. "Gov-ment men will be coming along that road in a day or two. They're going to move you, Granny, if you don't move yerself. Think a that. You gonna let some man, some strange man, do that to you?"

Granny looked at Geneva. A slow comprehension crept into her black eyes. "I ain't a-gonna die?"

"No, Granny. You ain't a-gonna die. The Lord ain't ready fer you yet. I think the Lord wants you to do something you ain't never done before. He's got more treasures fer you, earthly treasures."

"What ain't I never done before? I done *everything,* child. I'm old. I done it all."

"You ain't neither. You ain't done *nothing* but live on this farm and now you can't even die on this farm. Why, you ain't seen a moving picture, an' you ain't seen you an airplane except flying over the house, an you ain't seen . . . you ain't ever been to a foreign land, where you could a gone an' put some flowers on yer boy's grave over there from the war."

Granny screwed up her face. "You reckon the Lord'd want me to do *that*?"

"I don't know what the Lord wants you to do but I sure do know what he *don't* want you to do. He don't want you to die, that's fer sure!"

Granny sat back in her chair. *"Oh,* me, *ohhhhhh* my."* Her voice rose into an ancient keen. She pulled her apron over her face. She did not sob and there were no tears. Geneva, alarmed at the sounds from her granny, slipped off the porch. I'll give her time to wail, she decided. Geneva strolled toward the burying ground behind—behind what? The church house was gone. All that remained was the rocks of the foundation the women and children of long ago brought to their chosen place.

But the grass Daddy had planted after the grave movers had torn up the fabric of their lives was fresh and new. Under the brush she and Granny had piled over granddaddy's grave Geneva saw the first delicate lavender of a little blue flag. Only one flower bloomed, though the buds of the others were full.

Geneva picked it. Hearing silence from the back porch, she walked back to where Granny sat, calm now, head against the back of her chair, eyes closed.

"Here. I brought you this flower. Even though the tide is rising, this flower is growing. The blue flags er going to bloom one more time."

Granny studied the flower. She gently pulled the delicate tracing of a petal through thumb and finger. "You here with the mule an' you got a wagon?"

"I got the wagon. The mule's in the shed." Geneva answered.

"Well, bring it around. Load up my stuff. Let's go."

"Where, Granny? Where do you want to go?"

"I been studying what you been saying, child. I reckon I know what the Lord aims fer me to do. An' I'm gonna do it. Course, I'll need yer help. Yer daddy's a hard man to overcome."

Don't I know that, Geneva thought. "What can I do to help you, Granny?"

"You can bring that mule an' that wagon like I

done asked." Geneva was delighted to hear the sharpness in Granny's voice. "Bring it right up to the front door."

As Geneva pulled the mule and wagon up against the porch, Granny pushed her quilt chest through the front door. "Here. Put that in. An' my clock. Leave that old shucky mattress here. I'll get me a new one in town."

"In town?"

"If the Lord don't want me to die, then I gotta move. You say I should do something I ain't never done before. Well, I ain't never lived in town before. I guess I better hurry in an' find me a little place to live before all them farmers off the bottom get all the houses fer theirselves."

"You going to Bart's?"

"No. I'm going to my own house. First thing you gotta get in that wagon er all my treasures in blue jars. If we're going to town we'll need all them treasures. Go down in the cellar, gal. Bring up every blue jar you can find. There's peaches, an' green beans, an beet pickles, an' corn, an' treasures." Granny squinted at the sun. "If we're gonna get to town by dark we gotta get a move on."

"We?"

"You an' me. We can live in town. I kinda like living with you. Course, I don't want to live with you forever, but in town you can go to high school, an' you can make you a teacher. After you go off teaching, I don't want to live with you no more."

"Treasures in blue jars? Granny, what do you have in them blue jars besides dreams?"

"Money, child. Butter'n egg money. When yer granddaddy an' me run this farm the butter'n egg money was always mine. After them children were grown, I didn't hardly spend none. Money's in the blue jars, honey. What'd you think I been talking about?"

"But my dad don't want me to go. . . ."

"I 'spect he's used to it by now. You been gone a right smart while since they moved up on Toller's Ridge.

"Git a move on, gal. We got to get out of here if we're going. That sun's slipping down toward Tillet's Mountain."

About the Author

Though born and brought up in South Dakota, **NORMA COLE** spent many years in Michigan, where she raised a family and taught school. In 1984, she moved to Monticello, Kentucky, drawn by the quiet, rural life; the gentle, wooded mountains; and the wealth of stories to be listened to. Norma Cole lives in an old log house deep in the woods on Molehill Mountain and, along with writing for children, enjoys weaving bright, textured rugs. *The Final Tide* is based on extensive research and on stories she has heard from her neighbors.